AF838717

AERIAL ROBOTS - AERODYNAMICS, CONTROL AND APPLICATIONS

Edited by **Omar Dario Lopez Mejia**
and **Jaime Alberto Escobar Gomez**

Aerial Robots - Aerodynamics, Control and Applications
http://dx.doi.org/10.5772/65604
Edited by Omar Dario Lopez Mejia and Jaime Alberto Escobar Gomez

Contributors
Daniele Ventura, Andrea Bonifazi, Maria Flavia Gravina, Giandomenico Ardizzone, Esteban Valencia, Victor Hidalgo, Xavier Prats, Pablo Royo, Leonardo Camargo Forero, Mahmut Reyhanoglu, Andres Pérez, Sergio Preidikman, Bruno A. Roccia, Marcos L. Verstraete, Marcelo F. Valdéz, Balakumar Balachandran, Dean T. Mook, Antoni Grau, Rodrigo Munguia, Edmundo Guerra, Yolanda Bolea, Sarquis Urzua, Harijono Djojodihardjo

Published by InTech
Janeza Trdine 9, 51000 Rijeka, Croatia

© The Editor(s) and the Author(s) 2017
The moral rights of the editor(s) and the author(s) have been asserted.
All rights to the book as a whole are reserved by InTech. The book as a whole (compilation) cannot be reproduced, distributed or used for commercial or non-commercial purposes without InTech's written permission. Enquiries concerning the use of the book should be directed to InTech's rights and permissions department (permissions@intechopen.com).
Violations are liable to prosecution under the governing Copyright Law.

(cc) BY

Individual chapters of this publication are distributed under the terms of the Creative Commons Attribution 3.0 Unported License which permits commercial use, distribution and reproduction of the individual chapters, provided the original author(s) and source publication are appropriately acknowledged. More details and guidelines concerning content reuse and adaptation can be found at http://www.intechopen.com/copyright-policy.html.

Notice
Statements and opinions expressed in the chapters are these of the individual contributors and not necessarily those of the editors or publisher. No responsibility is accepted for the accuracy of information contained in the published chapters. The publisher assumes no responsibility for any damage or injury to persons or property arising out of the use of any materials, instructions, methods or ideas contained in the book.

Publishing Process Manager Viktorija Zgela
Technical Editor SPi Global
Cover InTech Design team

First published August, 2017
Printed in Croatia
Legal deposit, Croatia: National and University Library in Zagreb

Additional hard copies can be obtained from orders@intechopen.com

Aerial Robots - Aerodynamics, Control and Applications, Edited by Omar Dario Lopez Mejia and Jaime Alberto Escobar Gomez
p. cm.
Print ISBN 978-953-51-3463-3
Online ISBN 978-953-51-3464-0

We are IntechOpen, the world's leading publisher of Open Access books

Built by scientists, for scientists

3,500+
Open access books available

111,000+
International authors and editors

115M+
Downloads

151
Countries delivered to

Our authors are among the
Top 1%
most cited scientists

12.2%
Contributors from top 500 universities

WEB OF SCIENCE™

Selection of our books indexed in the Book Citation Index
in Web of Science™ Core Collection (BKCI)

Interested in publishing with us?
Contact book.department@intechopen.com

Numbers displayed above are based on latest data collected.
For more information visit www.intechopen.com

For EU product safety concerns:
IN TECH d.o.o., Prolaz Marije Krucifikse Kozulić 3, 51000 Rijeka, Croatia,
info@intechopen.com or visit our website at intechopen.com.

Meet the editors

Omar Lopez is an associate professor at the Department of Mechanical Engineering at the Universidad de los Andes in Bogota, Colombia. Prof. Lopez obtained his PhD degree in Mechanical Engineering from the University of Texas in Austin, USA, with specialization in thermal-fluid systems. His area of expertise is related to the simulation of external flows at moderate and high Reynolds number, including the simulation of the flow around MAVs, both fixed-wing and with rotors. Specifically, Prof. Lopez is interested in developing computational frameworks in which unsteady fluid dynamics simulation, structure interaction (RBD or flexible), and energy efficiency and control (including flow control) are highly coupled.

Jaime A. EscobarGomez received his bachelor's degree in Mechanical Engineering from the Universidad de America in Colombia and his master's degree in Aerospace Engineering from the Concordia University in Montreal, Canada. He is also a private pilot, an aviation maintenance technician, and an aviation aficionado. He joined the Department of Aerospace Engineering at the Universidad de San Buenaventura, Bogota, in 2008 where he has taught courses in aerodynamics, aircraft performance, flight dynamics, and case studies with applications in industry and CFD. His research work is focused on UAS and MAV design and innovation, applied low Reynolds number aerodynamics, and applied CFD for external flows. Mr. Escobar co-founded "Advector: Unmanned Systems" in 2010 where he leads the development team for the unmanned aerial systems.

Contents

Preface

Aerial robots (ARs) or unmanned aerial vehicles (UAVs) were initially developed for military applications such as surveillance, targeting, and strike missions. As the use of ARs has become more common in other civilian and industrial applications, the need of understanding all the technical elements and systems involved in these aircraft has become more evident for the academic community and the industry. Few years ago, the topic of aerial robots was exclusively related to the robotics community, so a great number of books about the dynamics and control of ARs and UAVs have been written and can be found in the literature. Indeed, the problem of control in a UAV is fundamental, but as the control technology for UAVs advances, the great interaction that exists between other systems and elements that are as important as control such as aerodynamics, energy efficiency, acoustics, structural integrity, and applications, among others has become evident. In this context, *Aerial Robots - Aerodynamics, Control, and Applications* is an attempt to bring some of these topics related to UAVs together in just one book and to look at a selection of the most relevant problems of UAVs in a broader engineering perspective.

This book (as suggested by its title) is divided into three sections: those contributed chapters related to aerodynamics and techniques to estimate aerodynamics properties for UAVs are grouped in the first section; the second section is devoted to control and dynamics of UAVs, and the final section gathers a choice of interesting civil applications of UAVs. In the aerodynamics section, the reader will find four contributed chapters related to the aerodynamics of UAVs with both fixed and rotary wings as well as flexible wings.

The estimation of aerodynamic properties such as forces and moments is the main topic of discussion, and several different techniques are exposed in order to make these estimations, ranging from analytical up to computational fluid dynamics (CFD). A computational framework to couple aerodynamics and elasticity (aeroelasticity) for UAVs with flexible wings is proposed in the first chapter. In this section, the use of distributed propulsion systems for flying wings and blended-wing-body UAV configurations with a potential improvement in energy and structure efficiency is also explored. The use of Coanda effect for lift enhancement in micro-aerial vehicles (MAVs) that use rotors is also discussed in this section as well as the aerodynamic analysis from three different approaches of the propeller of a production multirotor.

The second section of *Aerial Robots - Aerodynamics, Control, and Applications* is related to two very interesting topics: one related to nonlinear dynamics and control of ARs, which is a state-of-the-art topic. The second chapter of this section is related to a control that is specifically designed for an industrial application (i.e., sampling collection), which is also a new trend in control design for UAVs. Finally, the application section is devoted to two state-of-the-art problems in ARs: high-performance computing on board aerial robots and environmental monitoring using unmanned aerial systems (UAS).

It is expected that this book would serve as reference for graduate students as well as researchers in the field of aerial robots and unmanned aerial vehicles in the broad sense of an engineering problem and help as a bridge to reduce the gap between several different disciplines involved in this complex topic. Designers and engineers from the UAV industry may also benefit from the content of the book as it provides the state of the art in a variety of topics, and it may inspire the development of new commercial products. We hope that the way the book is organized helps the reader to find a specific topic of interest and to engage the reader as he/she goes from one section to the next one. Finally, I would like to acknowledge the contribution of Prof. Jonathan Camargo and Rodrigo Escobar for their collaboration during the review process of several chapters of this book.

Omar Dario Lopez Mejia
Associate Professor
Department of Mechanical Engineering
University of los Andes
Bogota, Colombia

Section 1

Aerodynamics

Chapter 1

Computational Aeroelasticity of Flying Robots with Flexible Wings

Sergio Preidikman, Bruno Antonio Roccia,
Marcos Leonardo Verstraete,
Marcelo Federico Valdez, Dean T. Mook and
Balakumar Balachandran

Additional information is available at the end of the chapter

http://dx.doi.org/10.5772/intechopen.69396

Abstract

A computational co-simulation framework for flying robots with flexible wings is presented. The authors combine a nonlinear aerodynamic model based on an extended version of the unsteady vortex-lattice method with a nonlinear structural model based on a segregated formulation of Lagrange's equations obtained with the Floating Frame of Reference formalism. The structural model construction allows for hybrid combinations of different models typically used with multibody systems such as models based on rigid-body dynamics, assumed-modes techniques, and finite-element methods. The aerodynamic model includes a simulation of leading-edge separation for large angles of attack. The governing differential-algebraic equations are solved simultaneously and interactively to obtain the structural response and the flow in the time domain. The integration is based on the fourth-order predictor-corrector method of Hamming with a procedure to stabilize the iteration. The findings are found to capture known nonlinear behavior of flapping-wing systems. The developed framework should be relevant for conducting aeroelastic studies on a wide variety of air vehicle systems.

Keywords: flexible wings, flapping wings, morphing wings, micro-air vehicles, aero-elasticity, co-simulation

1. Introduction

Natural flight is characterized by highly nonlinear and unsteady flows and wing-deformation patterns that include time-dependent in-and-out-of-plane bending, changing camber, spanwise

INTECH
open science | open minds

© 2017 The Author(s). Licensee InTech. This chapter is distributed under the terms of the Creative Commons Attribution License (http://creativecommons.org/licenses/by/3.0), which permits unrestricted use, distribution, and reproduction in any medium, provided the original work is properly cited. (cc) BY

twisting, expanding and contracting wing surfaces, and so on. A detailed description of the deformations of wings in different flight configurations is crucial for quantifying the influence of a wing's flexibility on the aerodynamic forces. Flight kinematics and aerodynamics have been identified and studied for a variety of insect and bird wings [1–13]. Advances in computational power have further enabled studies of highly flexible structures such as flapping and morphing wings of micro-air vehicles (MAVs). The primary components of numerical simulations are the aerodynamic solver, the structural-dynamic solver, and the communication between them. Early studies on fluid-structure interactions for vehicles with flapping wings have been based on either two-dimensional models or other simplified aerodynamic models based on thin-airfoil theory or unsteady panel methods or Euler methods and on linear beam finite elements [14–17].

Nakata and Liu [18] used an unsteady aerodynamic model based on a Navier-Stokes solver [19] coupled with a finite-element formulation specifically developed for insect flight that accounts for the distribution and anisotropy of the wing's veins and membranes. Chimakurthi et al. [20] proposed a computational framework, in which a nonlinear finite-shell-element solver is coupled with a Navier-Stokes flow solver [21]. Malhan et al. [22] analyzed flexible flapping wings by using a computational fluid dynamics-computational structural-dynamics (CFD-CSD) environment. They used a Reynolds-averaged Navier-Stokes (RANS) solver [23] along with an open-source multibody software [24]. Unger et al. [25] investigated the flow around a flexible airfoil based on a seagull's wing. They used a flow model based on the incompressible unsteady RANS equations (URANS) [26]. To model the structure and obtain the response, the authors generated a finite-element model that includes shells, solids, and contact elements for modeling the various structural members of a wing, and used an ANSYS [27] solver. Recently, Bose et al. [28] solved the incompressible two-dimensional flow field for a range of solid-to-fluid added-mass ratios (characteristic of natural flappers or MAVs) by using an arbitrary Lagrangian-Eulerian (ALE) formulation and a nonlinear discrete model of the structure. They coupled the models through a fourth-order Runge-Kutta scheme implemented in OpenFOAM [29], and found that the structural response experienced a supercritical Hopf bifurcation [30].

Computational difficulties and costs associated with the use of CFD models have led to alternative approaches. For example, unsteady vortex-lattice methods (UVLMs) are an effective and a simple choice [31–33]. Related to flapping wings, Taha et al. [34] identified five main contributors to flow characteristics during hover: the wing's translation and rotation, the leading-edge vortices (LEVs), wake capture, viscosity, and added-mass effects. UVLMs can be used to model all aspects except viscous and LEV effects. As shown by the experiments of Dickinson et al. [1], over the range of Reynolds numbers (75–4000) of hovering insects, viscous effects can be neglected; this makes the use of UVLMs suitable for modeling the aerodynamics of flapping wings and morphing wings.

The authors extend the approach proposed by Preidikman and Mook [32] and Preidikman [35]; a computational aeroelastic framework is presented. In these prior studies, the structures were much simpler than those considered here, and they were represented by means of traditional beam formulations based on assumed-modes/finite-element methods. In the authors' novel methodology, the dynamical system is partitioned into two subsystems (the structural and aerodynamic models) that communicate with each other across the boundary of the flow field

http://dx.doi.org/10.5772/intechopen.69396

(the surface of the structure) in a strong way. The computational environment can be considered a *strong co-simulation framework* [36]. Initial steps taken in this direction are presented in reference [37]. The underlying methodology is detailed in Section 2. The aerodynamic and structural models are presented in the third and fourth sections, respectively. The communication between the two models and the numerical integration scheme are presented in Section 5. In the next section, results obtained for systems with flapping and morphing wings are presented. Finally, concluding remarks are collected together and presented.

2. Methodology

The structural model (Simulator 1) and the aerodynamic model (Simulator 2) are different subsystems, which are constructed to exchange information bi-directionally in an iterative sequence that continuously improves the estimation of the aerodynamic loads (see **Figure 1**) and, consequently, the structure's response. The numerical scheme used by Simulator 2 is based on the UVLM, and the one used by Simulator 1 to solve the vehicle's equations of motion is based on Hamming's fourth-order predictor-corrector method [35]. This procedure was chosen for two reasons: (i) performance of Simulator 2 is better if the loads are evaluated at integral time steps and (ii) aerodynamic loads depend on the structure's acceleration. To combine the simulators, a technique for transferring information (TTI) based on radial basis functions is used.

The substantial deformations of the lifting surfaces induce significant changes in the aerodynamic loads, which in turn induce further changes in the deformation of the wings. This feedback between the aerodynamic loads and the movement of structural members (flowfield boundaries) generates strong coupling between the aerodynamic and structural models. In order to capture these complex interactions with a numerical model, a strong coupling method

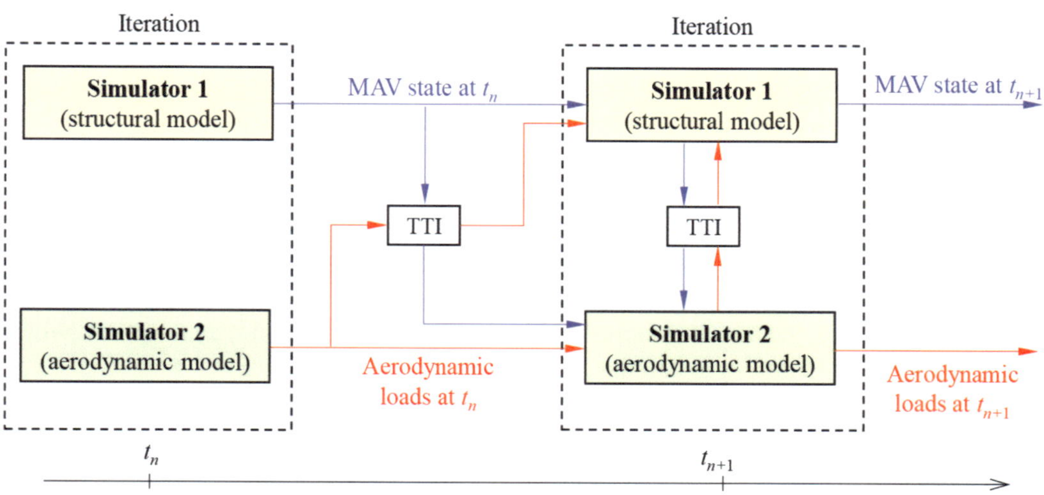

Figure 1. Schematic diagram of fluid-structure interaction framework [37].

is needed; the authors have chosen a so-called two-way non-monolithic strategy. Although Simulators 1 and 2 are computational implementations of independently modeled physical components, the coupling procedure is strong because information can be bi-directionally exchanged, and the chosen time step advance is the same for both models. Traditional approaches used to study unsteady aeroelastic characteristics may not be adequate for modeling the unusual fluid-structure interactions that characterize aerial vehicles with flexible flapping wings. Classical linear analyses can, in some instances, predict when aeroelastic systems become unstable, but they can neither detect subcritical instabilities nor describe post-instability motions. As the solutions are obtained in the time domain here, the authors have the capability for predicting both subcritical instabilities and post-instability motions [30].

3. Aerodynamic model

Here, UVLM is used to compute the aerodynamic loads. It can be applied to lifting surfaces of any planform, camber, and twist. The lifting surface may undergo any time-dependent deformation and execute any maneuver in moving air. The flow surrounding the lifting surface is assumed to be inviscid, incompressible, and irrotational over the entire flowfield, except at the solid boundaries and in the wakes. Due to the relative motion between the wing and the fluid and due to the viscous effects, vorticity is generated in the fluid in a thin region next to the wing's surface (the boundary layer). The boundary layers on the upper and lower surfaces are merged into a single vortex sheet. Vorticity in this sheet must be shed from the sharp edges (trailing edge, leading edge, and wing tips) so that the pressures are equal in the merging flows coming off the upper and lower sides of the lifting surface, the so-called Kutta condition. This shed vorticity forms the wake as it moves with the fluid in order for the pressure to remain continuous across the wake, which is force-free. The calculation of the velocity field associated with the vorticity is expedited by replacing the continuous vortex sheet with a lattice of discrete vortex lines. The boundary layers are represented by bound-vortex lattices, which are attached or bound to the lifting surface and move with it. The wakes are represented by free-vortex lattices, which move freely with the flow so that their position and vorticity distribution are determined as part of the solution [35, 37–39].

3.1. Mathematical formulation

The velocity of a fluid particle that occupies position, $\boldsymbol{r}$, at instant t is denoted by $\boldsymbol{V}(\boldsymbol{r};t)$. Since the flow is irrotational outside the boundary layers and wakes and considered to be incompressible, the velocity field can be expressed as the gradient of a scalar potential $\varphi(\boldsymbol{r};t)$, which is governed by the continuity equation:

$$\nabla^2\varphi(\boldsymbol{r};t) = 0. \tag{1}$$

In a three-dimensional flow, the velocity field satisfying Eq. (1) and associated with a straight, finite-length segment of a vortex line with circulation $\Gamma(t)$ is given by the Biot-Savart law:

$$V(\boldsymbol{r},t) = \frac{\Gamma(t)}{4\pi}\frac{\boldsymbol{L}\times\boldsymbol{r}_1}{\|\boldsymbol{L}\times\boldsymbol{r}_1\|_2^2 + (\delta\|\boldsymbol{L}\|_2)^2}[\boldsymbol{L}\cdot(\hat{\boldsymbol{e}}_1 - \hat{\boldsymbol{e}}_2)] \tag{2}$$

Here, $\boldsymbol{r}_1$ and $\boldsymbol{r}_2$ are the position vectors of the point of interest relative to the ends of the vortex segment; $\hat{\boldsymbol{e}}_1$ and $\hat{\boldsymbol{e}}_2$ are the unit vectors parallel to $\boldsymbol{r}_1$ and $\boldsymbol{r}_2$, respectively, and $\boldsymbol{L} = \boldsymbol{r}_1 - \boldsymbol{r}_2$ is the vector along the vortex segment. The term $\delta\|\boldsymbol{L}\|_2$ in Eq. (2) is added in order to avoid the singularity that appears when the point approaches the vortex line or its extension. The influence of the cutoff radius δ on the velocity is strongly felt in the immediate vicinity of the vortex line but is hardly noticeable elsewhere. A linear cutoff radius is another option [40].

3.2. Unsteady vortex-lattice method

In UVLM, the vortex sheets are replaced by lattices of short straight vortex segments with spatially constant circulation. These segments divide the wing surface into a finite number of typically nonplanar, quadrilateral elements of area with straight edges often called panels. The model is completed by joining the free-vortex lattices (wakes) to the bound-vortex lattice (lifting surface) along the edges where separation occurs. The separation location is user supplied and typically based on experience. Each quadrilateral element of the bound lattice has a single unknown circulation $G(t)$ instead of the four unknown circulations around each of the short, vortex-line segments along its edges. Consequently, the requirement of spatial conservation of circulation [41] is automatically satisfied throughout the lattices. Once all the $G(t)'$s are known, the $\Gamma(t)'$s around all of the straight vortex segments can be easily determined. In **Figure 2**, a representative example of the resulting bound and free-vortex sheets and vortex lattices after the discretization is shown.

The governing equation is complemented with the following boundary conditions: (i) the velocity field associated with the disturbance decays away from the body and its wakes. Hence,

$$\lim_{\|r\|_2\to\infty}\|\boldsymbol{V}_B(\boldsymbol{r};t) + \boldsymbol{V}_W(\boldsymbol{r};t) + \boldsymbol{V}_{LW}(\boldsymbol{r};t)\| = \|\boldsymbol{V}_\infty\|, \tag{3}$$

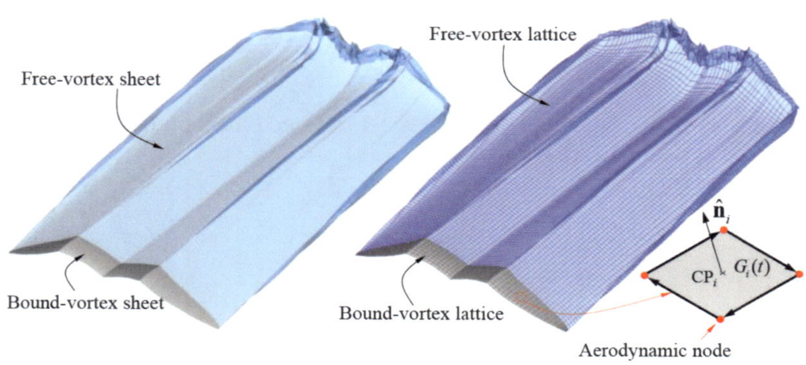

Figure 2. Vortex sheets and vortex lattices.

where $V_B(r;t)$, $V_W(r;t)$, and $V_{LW}(r;t)$ are the velocity fields associated with the bound-vortex lattice, the free-vortex lattice shed from the wing's trailing edge (including the tip), and leading edge, respectively, and V_∞ is the free-stream velocity. The velocity field obtained from the Biot-Savart law satisfies this condition. (ii) The normal component of the fluid velocity relative to the body's surface must be zero:

$$(V_\infty + V_B(r;t) + V_W(r;t) + V_{LW}(r;t) - V_P) \cdot \hat{n} = 0, \tag{4}$$

where V_P is the velocity of the body's surface, and $\hat{n}$ is a unit vector normal to the surface. Since there is a finite number of panels, and hence unknown circulations, Eq. (4) is only imposed at one point in each panel. Here, V_B is expressed in terms of the unknown values for the loop-vortex circulations $G_j(t)$ and the aerodynamic influence coefficients $A_{ij}(t)$ [35, 39]. To satisfy the unsteady Kutta condition at each time step, the vortex rings along the edges are shed into the flow where they have the same order as they had on the wing's surface. The vortex rings are moved downstream with the flow by moving the end points of their vortex segments, called nodes, with the local fluid-particle velocity V to new positions, denoted $r(t + \Delta t)$, according to the first-order approximation:

$$r(t + \Delta t) = r(t) + V(r, t)\ \Delta t. \tag{5}$$

3.3. Leading-edge separation model

Vortex shedding from the leading edge (LEV) depends on the angle between the local fluid velocity and the velocity of the edge of the moving wing's plane (the effective angle of attack). Several studies on conventional aircraft wings have reported that flow attached to the wing starts to separate when the effective angle of attack exceeds a critical value of 12–15°. Several numerical tools based on vortex-lattice methods account for LEV on highly swept delta wings [42]. Here, the effects of leading-edge separation are included through an on/off mechanism that consists mainly of computing the value of the effective angle of attack α_e at each time step and comparing it with a reference value α_c. Only when $\alpha_e \geq \alpha_c$ is leading-edge separation included [38].

3.4. Aerodynamic loads

The force on each element of the bound lattice is determined based on the pressure jump across the lifting surface at the control point; this calculation is carried out by using the unsteady Bernoulli's equation:

$$\partial_t \varphi(r, t) + \frac{1}{2} V(r;t) \cdot V(r;t) + \frac{p(r, t)}{\rho} = H(t) \tag{6}$$

Here, ∂_t denotes the partial time derivative at a fixed location in an inertial reference frame, ρ is the fluid density, p is the pressure, and $H(t)$ is the total energy per unit mass, which only depends on time and has the same value at every point in the whole domain of the flow [35].

4. Structural model

The equations governing the structure were developed for large rotations and displacements, called *primary motions*, and small/moderate rotations and displacements with respect to a moving reference frame, called *secondary motions*. The primary motions describe the position and orientation of each body as a whole, and the secondary motions describe deformations. The Floating Frame of Reference (FFR) formalism and Lagrange's method for constrained systems [43] were used to derive the equations of motion. The flying robot system is modeled as a collection of $\boldsymbol{n}_b$ rigid and deformable bodies subjected to $\boldsymbol{n}_c$ constraints, which are used to impose the connections and predefined motions. The fuselage (called the central body) is rigid, and the wings attached to it are assumed to be flexible. Three reference-point coordinates that typically coincide with either the body's center of mass or its centroid represent the rigid-body's position. The rigid body's orientation is described by Euler angles. In what follows, bold italic letters and bold letters are used for tensor notation and matrix notation, respectively.

4.1. Reference frames

In the FFR formalism, the rigid-body motion of the *k*th body is the primary motion, and the motion of points in the body's reference frame is the secondary motion. If the displacement field contains rigid-body modes, a set of conditions has to be imposed to define a unique displacement field with respect to the selected body reference. Consequently, $n_b + 1$ reference frames are used in order to describe the kinematics of the multibody system: (i) an inertial system $\mathbf{N} = \{\hat{\mathbf{n}}_1, \hat{\mathbf{n}}_2, \hat{\mathbf{n}}_3\}$ and (ii) a reference system fixed to each body k, $\mathbf{B}^k = \left\{\hat{\mathbf{b}}_1^k, \hat{\mathbf{b}}_2^k, \hat{\mathbf{b}}_3^k\right\}$. The set of vectors $\hat{\mathbf{n}}_i$ and $\hat{\mathbf{b}}_i^k$ for $i = 1, 2, 3$ form a dextral orthonormal basis.

4.2. Velocity for an arbitrary point in a body

Let us consider the *k*th flexible body, which moves and deforms in space. The components of the absolute translational velocity at an arbitrary point in the body P, $\dot{\boldsymbol{R}}_{p}^k$, can be expressed in the segregated form:

$$\dot{\boldsymbol{R}}_p^k = \dot{\boldsymbol{R}}^k + Q_{\mathbf{NB}^k}\left[\dot{\boldsymbol{r}}_0^k + \dot{\boldsymbol{u}}^k + \boldsymbol{\omega}^k \times \left(\boldsymbol{r}_0^k + \boldsymbol{u}^k\right)\right], \quad \text{for } k = 1, \ldots, n_b \tag{7}$$

Here, $\dot{R}^k$ is the velocity of the origin of the body-fixed frame $\mathbf{B}^k$; r_0^k is the position vector of an arbitrary point P on the *k*th body in the undeformed configuration, and $\dot{\boldsymbol{r}}_0^k$ is identically zero because the differentiation is performed with respect to the $\boldsymbol{B}^k$ frame; $\boldsymbol{u}^k$ is the elastic displacement vector of the point P and $\dot{\boldsymbol{u}}^k$ represents its time derivative with respect to the body-fixed frame; $\boldsymbol{\omega}^k$ is the angular velocity vector of $\mathbf{B}^k$ relative to the inertial frame $\mathbf{N}$; and $Q_{\mathbf{NB}^k} : \mathbf{B}^k \rightarrow \mathbf{N}$ represents a rotation tensor from the body-fixed frame $\mathbf{B}^k$ to the inertial frame $\mathbf{N}$. After algebraic manipulations, Eq. (7) can be written as

$$\dot{R}_p^k = \dot{R}^k + Q_{\mathbf{NB}^k}\dot{u}^k - Q_{\mathbf{NB}^k}\left[\left(\overline{r}_0^k + \overline{u}^k\right)\omega^k\right], \quad \text{for } k = 1, \ldots, n_b, \tag{8}$$

where $\overline{r}_0^k$ and $\overline{u}^k$ are skew-symmetric tensors associated with r_0^k and u^k, respectively. Their actions on vectors are equivalent to the cross products. The tensor $Q_{\mathbf{NB}^k}$ is a function of $\mathbf{q}^k$, the vector of absolute coordinates that describes the motion of the Floating Frame of Reference $\mathbf{B}^k$ attached to body k, or primary motions. The vector $\mathbf{q}^k$ is given by

$$\mathbf{q}^k = \left[\left(\mathbf{R}^k\right)^T \left(\boldsymbol{\theta}^k\right)^T\right]^T, \tag{9}$$

where $\mathbf{R}^k$ is a set of coordinates for the origin of the body reference frame $\mathbf{B}^k$, and $\boldsymbol{\theta}^k$ is a set of rotational parameters that orient the body reference frame (e.g., Euler angles, Rodriguez parameters, etc.). When high-order terms are not considered, the elastic displacement vector solved in the body-fixed frame is $\mathbf{u}^k = \mathbf{N}^k\mathbf{p}^k$, where $\mathbf{N}^k$ is a matrix of shape functions for interpolating the elastic displacement across the body domain, and $\mathbf{p}^k$ is a time-dependent vector of generalized coordinates that is used to describe secondary motions.

4.3. Equations of motion

Once the absolute vector of translational velocity for any point on the body is known, the equations of motion can be derived. Additionally, it is important to consider that a flexible body can be constrained [43, 44], by either linking it with other bodies or by choosing the parameters to describe primary motions; only holonomic constraints are considered. To simplify the manipulation, Lagrange's equations are separated into two groups: one corresponding to $\mathbf{q}^k$ and the second to $\mathbf{p}^k$. The constrained Lagrange's equations are given by

$$d_t\left(\partial_{\dot{\mathbf{q}}^k}T^k\right) - \partial_{\mathbf{q}^k}T^k + \boldsymbol{B}_{\mathbf{q}^k}^T\boldsymbol{\lambda}^k = \left(\boldsymbol{Q}_q^k\right)^T, \quad \text{and} \quad d_t\left(\partial_{\dot{\mathbf{p}}^k}T^k\right) - \partial_{\mathbf{p}^k}T^k + \boldsymbol{B}_{\mathbf{p}^k}^T\boldsymbol{\lambda}^k + \partial_{\mathbf{p}^k}U^k = \left(\boldsymbol{Q}_p^k\right)^T, \tag{10}$$

which are complemented by a set of algebraic-constraint equations expressed as

$$\boldsymbol{\Phi}^k\left(\mathbf{q}^{k-1}, \mathbf{q}^k, \mathbf{q}^{k+1}; t\right) = \mathbf{0}. \tag{11}$$

Here, the $\mathbf{q}^k\left(\mathbf{p}^k\right)$ were defined previously, and $d_t, \partial_{\mathbf{q}^k}\left(\partial_{\mathbf{p}^k}\right)$ as well as $\partial_{\dot{\mathbf{q}}^k}\left(\partial_{\dot{\mathbf{p}}^k}\right)$ denote time derivatives, the partial derivatives with respect to $\mathbf{q}^k\left(\mathbf{p}^k\right)$ and with respect to $\dot{\mathbf{q}}^k\left(\dot{\mathbf{p}}^k\right)$, respectively. $\boldsymbol{\lambda}^k$ is the vector of Lagrange multipliers, $\boldsymbol{B}_{\mathbf{q}^k}\left(\boldsymbol{B}_{\mathbf{p}^k}\right)$ is the Jacobian tensor of constraints associated with the coordinates $\mathbf{q}^k\left(\mathbf{p}^k\right)$, T^k is the kinetic energy of the k-th body, U^k is the elastic potential energy, and $\mathbf{Q}_q^k\left(\mathbf{Q}_p^k\right)$ is the generalized load vector corresponding to the vector $\mathbf{q}^k\left(\mathbf{p}^k\right)$. The kinetic energy T^k follows from $\dot{\boldsymbol{R}}_p^k$ in Eq. (8), and then the equations for *primary* and *secondary motions* for the k-th body are as follows:

$$\mathbf{M}^k\ddot{\mathbf{q}}^k + (\mathfrak{M}^k)^T\ddot{\mathbf{p}}^k + \mathbf{B}_{\mathbf{q}^k}^T\lambda_k = (\mathbf{Q}_q^k)^T,$$
$$\mathbf{m}^k\ddot{\mathbf{p}}^k + \mathfrak{M}^k\ddot{\mathbf{q}}^k + \mathbf{K}^k\mathbf{p}^k + \mathbf{B}_{\mathbf{p}^k}^T\lambda_k = (\mathbf{Q}_p^k)^T, \quad (12)$$
$$\mathbf{\Phi}^k(\mathbf{q}^{k-1}, \mathbf{q}^k, \mathbf{q}^{k+1}, \mathbf{p}^{k-1}, \mathbf{p}^k, \mathbf{p}^{k+1}; t) = \mathbf{0}, \quad \text{for } k = 1, \ldots, n_b.$$

Here, $\mathbf{M}^k$ is the mass matrix for primary motions, which is differentiable, symmetric, and at least positive-semi-definite; $\mathfrak{M}^k$ is the mass matrix that couples primary and secondary motions; $\mathbf{m}^k$ is the metric tensor for secondary motions, which is constant, symmetric, positive-definite, and its matrix representation is the "elastic mass matrix"; $\mathbf{K}^k$ is the stiffness matrix for the secondary motions; and $\mathbf{\Phi}^k$ is the set of holonomic-rheonomic constraint equations associated with the k-th body. It should be noted that $\mathbf{q}^T = \left\{(\mathbf{q}^1)^T, \ldots, (\mathbf{q}^m)^T\right\}$ for $m \le n_b$ and $\mathbf{p}^T = \left\{(\mathbf{p}^1)^T, \ldots, (\mathbf{p}^n)^T\right\}$ for $n \le n_b$.. The total number of coordinates is calculated through $6n_b + \sum_{k=1}^{n_b} N_k = n_{\text{coord}}$ (N_k is the number of elastic generalized coordinates of the k-th body). Thus, the number of degrees of freedom is $n_{\text{dof}} = n_{\text{coord}} - n_c$. A beam element based on Euler-Bernoulli/Rayleigh theory is used. Hermite polynomials are used to interpolate displacement/rotation fields in each finite element from nodal values. First-order polynomials are used to interpolate elongation and torsion, and third-order polynomials are used to interpolate bending [45]. Finally, the equations of motion for the complete multibody system are obtained by assembling the equations of motion for each body:

$$\mathbf{M}\ddot{\mathbf{x}} + \mathbf{B}_{\mathbf{x}}^T\lambda = \mathbf{F},$$
$$\mathbf{\Phi} = \mathbf{0}. \quad (13)$$

Here, $\mathbf{M} \in \mathbb{R}^{n_{\text{coord}} \times n_{\text{coord}}}$ is the global mass matrix, $\mathbf{B}_{\mathbf{x}} \in \mathbb{R}^{n_{\text{coord}} \times n_c}$ the constraint Jacobian matrix, $\mathbf{x} \in \mathbb{R}^{n_{\text{coord}} \times 1}$ is the global vector of generalized coordinates, $\boldsymbol{\lambda} \in \mathbb{R}^{n_c \times 1}$ is the global vector of Lagrange's multipliers, $\mathbf{F} \in \mathbb{R}^{n_{\text{coord}} \times 1}$ is the global vector of forces, and $\mathbf{\Phi} \in \mathbb{R}^{n_c \times 1}$ is the set of all constraints for the multibody system.

5. Communication between models and numerical integration

In this section, the schemes for exchanging information (loads, displacements, and velocities) between the structural and aerodynamic models and integrating the resulting algebraic-differential equation are described.

5.1. Information transfer between simulators for aerodynamics and structural dynamics

At each time step, the aerodynamic and structural models exchange information bi-directionally in an iterative sequence to improve the estimates of both the structure's response and the aerodynamic loads (see **Figure 3**). To this end, the authors propose the use of an interpolation procedure based on radial basis functions (RBFs) [46].

A linear relationship between the generalized displacements associated with both the aerodynamic mesh and the structural mesh is established through the following coupling matrix:

$$\mathbf{w} = \mathbf{G}_{AE}^{NA}\ \mathbf{u}, \qquad \text{and} \qquad \mathbf{v} = \mathbf{G}_{AE}^{CP}\ \mathbf{u} \tag{14}$$

Here, $\mathbf{w}$ and $\mathbf{v}$ are the displacements of the nodes and control points of the aerodynamic mesh, and $\mathbf{u}$ is the generalized nodal displacements of the structural mesh. $\mathbf{G}_{AE}^{NA}$ is the coupling matrix that relates $\mathbf{w}$ with $\mathbf{u}$, and $\mathbf{G}_{AE}^{CP}$ relates the control-point and the structural-node displacements. As a first step, a continuous scalar function $s(\mathbf{x})$ is generated by using the radial basis functions $\phi(|\mathbf{x} - \mathbf{x}_i^e|)$ [47, 48], where x is the location for the evaluation of s, and $\mathbf{x}_i^e$ is the positions of centers (structural nodes in the present case). It is required that the function s recovers the generalized structural displacements when it is evaluated at the centers. For the component z of $\mathbf{u}$ (denoted by u_i^z),

$$u_i^z = s(\mathbf{x}_i^e) = \alpha_1\ \phi(|\mathbf{x}_i^e - \mathbf{x}_1^e|) + \cdots + \alpha_i\ \phi(|\mathbf{x}_i^e - \mathbf{x}_i^e|) + \cdots + \alpha_{ne}\ \phi(|\mathbf{x}_i^e - \mathbf{x}_{ne}^e|) \qquad for \quad i = 1, 2, \ldots, n_e. \tag{15}$$

This constitutes a set of n_e (number of structural nodes) linear algebraic equations, which have the matrix form $\mathbf{A}_{ee}\boldsymbol{\alpha} = \mathbf{u}^z$, where $\mathbf{A}_{ee} \in \mathbb{R}^{n_e \times n_e}$ contains the functions φ evaluated at the structural nodal positions, $\boldsymbol{\alpha}$ is a column vector that contains the coefficients α_i, and $\mathbf{u}^z$ is an array which collects the z-components associated with the nodal translations of the structural mesh.

Solving for the unknown $\boldsymbol{\alpha}$, makes it possible to approximate the aerodynamic nodal displacements by evaluating the function $s(\mathbf{x})$ at the aerodynamic nodal positions $\mathbf{x}_i^a$. Again, this

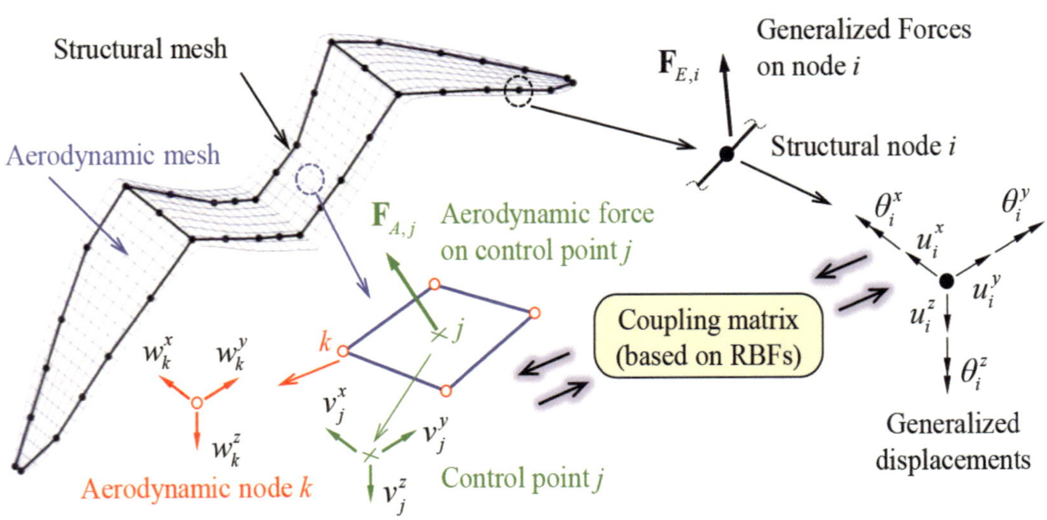

Figure 3. Schematic diagram of transfer technique between simulators.

procedure leads to a linear algebraic system $\mathbf{A}_{ae}\boldsymbol{\alpha} = \mathbf{w}^z$, where $\mathbf{A}_{ae} \in \mathbb{R}^{n_a \times n_e}$ (n_a is the number of aerodynamic nodes) contains the functions ϕ evaluated at the aerodynamic nodal positions. Combining the aforementioned two linear systems, a direct relation between $\mathbf{w}^z$ and $\mathbf{u}^z$ is obtained, $\mathbf{w}^z = \mathbf{H}\,\mathbf{u}^z$, where $\mathbf{H} = \mathbf{A}_{ae}\mathbf{A}_{ee}^{-1}$. If the same RBFs are used to interpolate the $x(\mathbf{w}^x)$ and $y(\mathbf{w}^y)$ components of $\mathbf{w}$, a similar relation is found. Therefore,

$$\mathbf{w}^r = \mathbf{H}\mathbf{u}^r, \quad \text{for } r = x, y, z. \tag{16}$$

In fact, matrix $\mathbf{H}$ is a linear map that relates the aerodynamic and structural nodal displacements. According to prior studies [35, 49], the procedure to obtain each one of the matrices introduced above consists of the following: (i) computing $\mathbf{H}$ based on the data associated with the structural and aerodynamic meshes, (ii) obtaining $\mathbf{G}_{AE}^{NA}$ from $\mathbf{H}$, and (iii) computing $\mathbf{G}_{AE}^{CP}$ [49]. The last matrix is used to transform the aerodynamic loads $\mathbf{F}_A$, acting at the control points of the aerodynamic grid, to an equivalent set of forces $\mathbf{F}_E$ acting on the finite-element mesh, $\mathbf{F}_E = \left(\mathbf{G}_{AE}^{CP}\right)^T \mathbf{F}_A$. The structural-aerodynamic force relation is obtained by establishing some kind of abstract equivalence; that is, both systems of forces must perform the same virtual work for any given virtual displacement [35]. In particular, this procedure has been developed for combining UVLM grids with structural meshes based on beam finite elements. As the coupling matrix only depends on the kind of RBFs used and the clouds of points to be interpolated, the same formulation is valid for other types of finite-element meshes.

5.2. Numerical integration of the index-1 DAEs

The set of equations of motion for the entire multibody system represents an index-3 system of DAEs. These dynamic equations are, in general, nonlinear. To solve them by means of standard solvers for ordinary differential equations (ODEs), an index reduction is required for the set of DAEs [44]. The methodology adopted in this work includes differentiation of the constraint equations twice with respect to time. This new set of equations is often called constraint acceleration level and is given by

$$\ddot{\boldsymbol{\Phi}}(\mathbf{x}, \dot{\mathbf{x}}; t) = \mathbf{B}_\mathbf{x}\ddot{\mathbf{x}} + 2\partial_\mathbf{x}(\partial_t\boldsymbol{\Phi})\dot{\mathbf{x}} + \partial_{tt}\Phi + \partial_\mathbf{x}(\mathbf{B}_\mathbf{x}\dot{\mathbf{x}})\dot{\mathbf{x}} = \mathbf{0}, \tag{17}$$

where $\mathbf{x}^T = \left\{\mathbf{q}^T, \mathbf{p}^T\right\}$, $\partial_\mathbf{x}$ denotes the partial derivative with respect to the vector $\mathbf{x}$, and ∂_{tt} denotes the second partial time derivative. The equations of motion for the multibody system (13) can be rewritten together with the acceleration level constraint equations, as an index-1 system of DAEs as follows:

$$\begin{bmatrix} \mathbf{M} & \mathbf{B}_\mathbf{x}^T \\ \mathbf{B}_\mathbf{x} & \mathbf{0} \end{bmatrix} \begin{Bmatrix} \ddot{\mathbf{x}} \\ \boldsymbol{\nu} \end{Bmatrix} = \begin{Bmatrix} \mathbf{Q} \\ \boldsymbol{\kappa} \end{Bmatrix}, \tag{18}$$

where $\boldsymbol{\kappa} = -2\partial_\mathbf{x}(\partial_t\boldsymbol{\Phi})\dot{\mathbf{x}} - \partial_{tt}\boldsymbol{\Phi} - \partial_\mathbf{x}(\mathbf{B}_\mathbf{x}\dot{\mathbf{x}})\dot{\mathbf{x}}$.

Numerical integration of Eq. (18) is susceptible to instabilities as a consequence of truncation procedures and round-off errors. The evident one is that the position and velocity constraints are no longer exactly satisfied; that is, a drift-out of the constraints does exist. In the literature, several stabilization methods to correct this numerical drift can be found, among which the most widely used due to its simplicity is Baumgarte's technique [50]. Another technique currently used consists in discretizing numerically the ODE and projecting the approximate solution onto the selected constraints manifold (coordinate projection) [51]. In this work, the authors have adopted the coordinate projection method to control/eliminate the numerical drift. As mentioned in Section 2, the numerical scheme used by Simulator 2 is well known and can be found in prior work [35, 39]. On the other hand, the numerical procedure adopted for Simulator 1 to solve the equations of motion of the aerial robot is based on Hamming's fourth-order predictor-corrector method [35, 52].

5.2.1. Co-simulation strategy

During a time step Δt, the wakes are convected to their new positions with the local fluid-particle velocity. Simultaneously, the structure of the aerial robot moves to its new position as a result of the acting forces and constraint equations. This concept is implemented by performing the following sequence of steps to calculate the solution at time $t + \Delta t$ as follows:

1. Simulator 2 is used to predict the new position of the wakes. A fluid particle in the wake moves from its current position $r(t)$ to its new position $r(t + \Delta t)$ according to Eq. (5). During the rest of the procedure for this time step, the wake is frozen.

2. The current loads computed by Simulator 2 are used in Simulator 1 to predict the response of the structure.

3. The current state of the structure is used as input to Simulator 2 and the loads are recalculated. Then, these loads are used as input to Simulator 1 and the state of the structure is updated again. This step is repeated until convergence. Usually, three to seven iterations are required to reduce the error to less than 10^{-10}.

4. The final position and velocity of the structure are evaluated by using Simulator 1, and used by Simulator 2 to recalculate the flow field and obtain the final estimate for the aerodynamic loads.

The procedure described above needs information from four previous time steps. At the beginning of the procedure, this information does not exist; so, the authors have used a special starting scheme: at $t = 0$, the initial conditions are used by Simulator 2 to calculate the aerodynamic loads ignoring the contribution of $\partial_t \varphi$. Further studies on this issue are needed to definitively state that $\partial_t \varphi$ can be neglected at $t = 0$.

6. Numerical results

In this section, results obtained with the implementation of the proposed methodology written in Fortran 90 are presented. The implementation is structured in a modular organization; that is, so that each part can be individually removed and replaced, as well as the capability of adding new models without modifying the general structure of the program. Although the

http://dx.doi.org/10.5772/intechopen.69396

parallelization of the code is not implemented, acceptable computing times, for full nonlinear simulations, were achieved. Simulations involving only structural models may take only a few minutes, but aeroelastic simulations including unsteady, nonlinear aerodynamics may take above 1 day on average in a desktop computer with an i7 processor, and RAM DDR3 of 4 GB.

6.1. Aerial vehicles with flapping wings

In this subsection, a series of numerical results obtained with the developed computational tool for a mechanical system based on flapping wings is presented. The first set of results consists of the aerodynamic analysis of MAVs with flapping wings, including some validations against experimental data. The second set of results is related to the study of the unsteady aerodynamics and nonlinear dynamics of MAVs, which explains some basic features of the involved flight mechanisms.

6.1.1. Aerodynamics of flapping wings

In this first study, the lift forces obtained from numerical simulations and comparisons of them with the experimental data reported by Dickinson et al. [1] are presented. The dynamically scaled experimental model imitated a *Drosophila melanogaster*, dubbed Robofly, with a wing span of 25 cm (from the force sensor to the wingtip). The wing executed an insect-like flapping motion at a frequency of 0.145 Hz with the wing tip tracing out a figure of eight. The viscosity of the oil, the length of the wing, and the flapping frequency were chosen in order to match the Reynolds number (*Re*) typical of the flight of a fruit fly ($Re = 136$). The kinematic pattern employed by Dickinson's team consisted of a stroke amplitude of 160°, and an angle of attack at midstroke of 40° for both upstroke and downstroke. In **Figure 4**, the authors show the three kinematic patterns used: (i) wing rotation precedes the reversal stroke by 8% of the wing-beat cycle (advanced pattern), (ii) wing rotation occurs symmetrically with respect to the reversal stroke (symmetrical pattern), and (iii) wing rotation is delayed with respect to the stroke reversal by 8% of the stroke cycle (delayed pattern). Details of the kinematic model used in this work to prescribe the motion of the Robofly's wing, as well as an extensive study of kinematical parameters can be found in reference [53].

In **Figure 5**, numerical results for the lift force (advanced pattern) with and without leading-edge separation and the comparison with those obtained by Dickinson et al. are shown. The results obtained are very encouraging because they show better agreement than those reported in previously published comparisons such as the CFD study by Sun and Tang [54], and the two-dimensional aerodynamic model developed by Ansari et al. [55, 56]. These results are significant because they justify the use of the nonlinear unsteady vortex-lattice method to study the three-dimensional aerodynamics of insects executing different maneuvers. Similar results were found for the symmetrical and delayed patterns [38]. These results suggest that the present model will also serve well for the study of aerial robots.

The Fast Fourier Transforms (FFTs) of the force data from numerical simulations and experiments were also computed and the results are shown in **Figure 5b**. The frequency content for the experimental data clearly shows the flapping frequency of the motion (point A in **Figure 5b**, $n_f = 0.1446$ Hz) and twice this frequency (point B in **Figure 5b**, $2n_f = 0.2893$ Hz) along with a

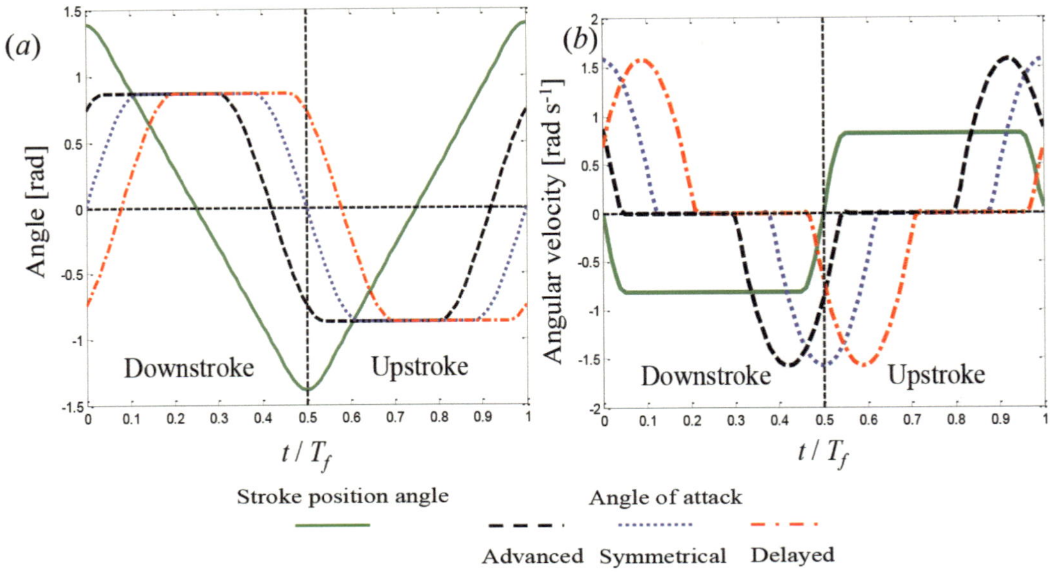

Figure 4. (a) Stroke position angle and angle of rotation of the wing for three different rotational timings. (b) Time derivatives of the stroke position angle and the rotation angle for three rotational timings. Here, T_f is the flapping period.

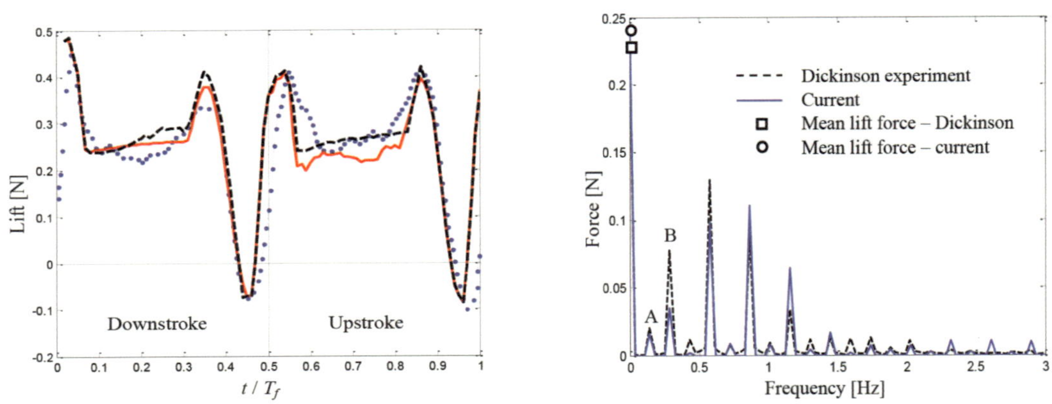

Figure 5. Comparison of numerical results with experimental data for the advanced pattern (first-stroke cycle). Dotted line for Dickinson's data, solid line for numerical results without LEV, and broken line for numerical results with LEV.

number of harmonics. These harmonics appear because there are two half-strokes per wing-beat cycle (the wing-passing frequency). The frequency content of the predicted lift force closely matches the flapping frequency of the motion ($n_f = 0.145$ Hz and $2n_f = 0.29$ Hz). Another indication of the quality of the numerical prediction is given by the mean lift, $\overline{L}$. The square and circular symbols in **Figure 5b** represent the experimental and predicted mean lift force, respectively. The difference between the predicted mean lift force and the experimental mean lift force for the advanced pattern is about 4.5%. For a detailed study of the aerodynamic of flapping wings using a modified version of the unsteady vortex-lattice method, the reader is referred to the work of Roccia et al. [38].

6.1.2. *Dynamics of flapping wings*

In this section, the numerical results obtained for the dynamic behavior of a fruit fly (*D. melanogaster*) in hovering flight are presented. The general equations of motion were adapted to consider only imposed deformations. These include torsion and bending in the directions normal and parallel to the wing's chord. Details of the kinematics and the dynamic models that allow one to prescribe such patterns of deformation are given in reference [57, 58]. Data reported by Fry et al. [58] on the actual kinematics of a fruit fly in hover were used to describe the wing motion over a flapping cycle (see **Figure 6a** where the solid blue line represents the stroke deviation angle, the dotted red line represents the stroke position angle, and the broken black line represents the rotation angle). Due to the complex motion experienced by the wings during a stroke cycle, the wake shed from the leading edge during the downstroke is likely cut by the wing during the upstroke. A wake rupture model is not yet available, and therefore no leading-edge separation was allowed in this case. For this numerical experiment, the following parameters were chosen: flapping frequency $n_f = 210$ Hz; wing span $R = 2.5$ mm and wing area $S = 2.21$ mm^2; air density $\rho_{air} = 1.2$ kg/m^3; insect's mass of 0.84 mg. A spatial discretization of the MAV of 1405 aerodynamic panels was used: 200 panels over each of the wings and 1005 panels over the central body (head, thorax, and abdomen of the insect). Only the first stroke cycle was simulated, which was discretized into 100 time steps. In **Figure 6b**, the authors show the computational model of the insect where the relative size and topology of the aerodynamic grid can be appreciated.

Two cases are presented next. In the first, the authors consider a rigid wing, and in the second, the authors consider a flexible wing with prescribed deformation. The prescribed deformation consists of wing torsion, which varies linearly along the wing span, combined with bending in the normal and tangential direction of the wing chord. The values used for the three different deformations are based on a detailed study carried out by the authors on the influence of

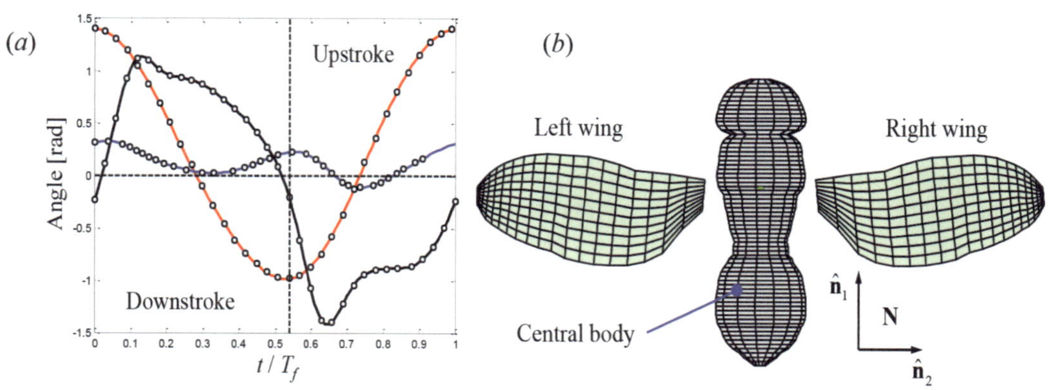

Figure 6. (a) Actual kinematics of a fruit fly in hovering, circular markers indicate experimental data. (b) Computational model of the insect.

spanwise twisting and bending on the lift-force generation in flapping-wing MAVs [58]. In order to show the role of the wing's flexibility in the production of lift, the parameters that regulate the deformation were tuned in order to increase the lift force throughout the stroke cycle with respect to the rigid-wing model. The set of initial conditions for the two cases studied consists of a body angle of 75° and a stroke plane angle of 15°. These values produce a horizontal stroke plane. Both the linear and angular velocities of the central body are zero at $t = t_0$.

In **Figure 7**, the authors show the lift and horizontal forces obtained numerically for both the rigid-wing model and the flexible-wing model along with experimental measurements obtained by Fry et al. [58] for a *Drosophila* in hover. The most important requirement of hovering flight is undoubtedly that the value of the vertical force must compensate for the insect's weight. In **Figure 7a**, a peak force can be observed at the center of each half-stroke, mainly during the upstroke. The peak force produced by the flexible wing is approximately 30% higher than what the rigid-wing model predicts. This fact is also reflected in **Figure 7d**, in which the degree of freedom associated with the vertical displacement of the center of mass of the central body has been plotted. When the authors use a rigid-wing model, the lift force produced can hardly support the insect's weight, showing a slight downward movement at the end of the stroke cycle. On the other hand, when the authors impose a suitable deformation pattern, not only the lift generated is sufficient to balance the insect's weight, but it also rises almost continuously.

In order to assess the quality of numerical simulations, the aerodynamic forces obtained numerically are compared with experimental measurements. As can be observed in **Figure 7a**, the time evolution of the lift force for the rigid and flexible wing shows a shape and trend similar to the curve reported by Fry et al. [58]. Nevertheless, lift levels predicted by the numerical model are lower than those measured experimentally (approximately 14% lower for flexible wing case). Possibly, this fact is because the LEV phenomenon was not taken into account. On the other hand, the predicted horizontal force (**Figure 7b**) departs quite noticeably from the results of Fry et al. However, the shape and trend is well captured. It should be noted that the aerodynamic model used is inviscid. Furthermore, UVLM does not account for leading-edge suction. In addition, the horizontal force is always opposite to the wing's motion, leading to a longitudinal oscillation of the insect's body. This dynamic behavior can be observed in **Figure 7c**.

Unlike forward thrust and lift, sideways thrust cancels instantly due to the bilateral symmetry of the wing motion (see **Figure 7f**). Similarly, because each wing contributes to yaw and roll torque in opposite directions, these moments cancel out. On the other hand, the sign and magnitude of pitch torque is the same for both wings. This fact directly affects the pitch angle (see **Figure 7e**). Fry et al. [58] observed that pitch torque averages zero over a complete cycle, and therefore the insect's body experiences an oscillation in the pitch angle. In real insects, this phenomenon is controlled by the wing kinematics, the flapping frequency, and the precise motion of the head and abdomen relative to the insect's thorax. Such a control mechanism is essential to prevent the pitch torque from growing indefinitely. In this work, no control strategy was applied, and therefore the pitch angle does not behave as expected. It decreases continuously for most of the flapping cycle, only showing a slight increase toward the end of the upstroke.

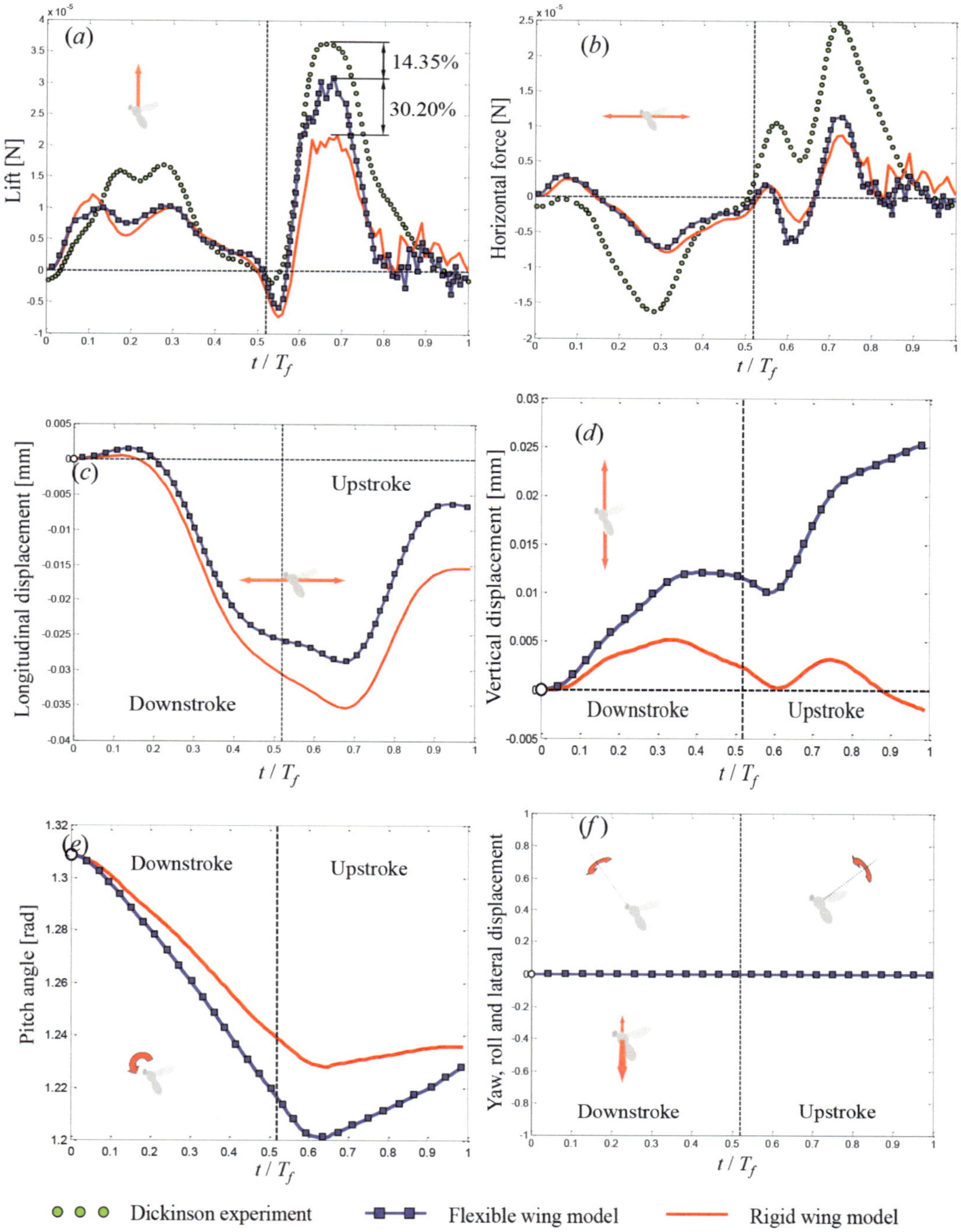

Figure 7. Aerodynamic forces. (a) Lift force. (b) Horizontal force. (c) Longitudinal displacement. (d) Vertical displacement. (e) Pitch angle. (f) Roll, yaw, and lateral displacements.

6.2. Aerial vehicles based on the morphing-wing concept

In this section, a series of numerical results for an aerial vehicle with morphing wings is presented. The first set describes the aerodynamic behavior of a bio-inspired, morphing-wing concept. The second set of results is related to the aeroelastic study of an aerial-robot wing model whose planform geometry is based on the seagull wing.

6.2.1. Aerodynamics of morphing wings

The model's geometry preserves certain morphological parameters of a gull wing [59], and a folding wing mechanism is used to change the wing's shape. The model consists of the right and the left wings, which are joined at the wing root. Because the two parts are mirror images, the model is represented by only the right wing (see **Figure 8a**).

Each wing is represented by two parts: the inner wing and the outer wing. In **Figure 8a**, the boundary between these parts is pointed out. A seagull-wing profile obtained by Liu et al. [60] is used. The center line, contained in the plane of symmetry *xz*, does not move during the morphing process. The inner and outer parts of the wing are connected and can move with respect to each other. The folding motion of the inner and outer wings is described by the dihedral angles θ_B and θ_C, respectively (see **Figure 8b** and **c**). During the simulation, the wings remain fully extended until a time T_I, and then the morphing process begins by varying the dihedral angles according to expression (19) until a time T_S, where the wings take a configuration that is maintained in a steady state and is determined by the angles θ_B^S and θ_C^S. The simulation ends at a time T_F.

$$\theta_B(t) = \theta_B^S\left(\frac{t-T_I}{\Delta T} - \frac{1}{2\pi}\sin\left(\frac{2\pi(t-T_I)}{\Delta T}\right)\right) \text{ and } \theta_C(t) = \theta_C^S\left(\frac{t-T_I}{\Delta T} - \frac{1}{2\pi}\sin\left(\frac{2\pi(t-T_I)}{\Delta T}\right)\right). \quad (19)$$

Here $\Delta T = T_S - T_I$. Three passages through the morphing processes were simulated with a mesh of 8×32 panels in the chordwise and spanwise directions, respectively. The morphing wing was reconfigured in time by means of Eq. (19) with the values given in **Table 1**. In each case, two conditions were considered: (i) with shedding vorticity from the wing tips and (ii) without shedding vorticity from the wing tips.

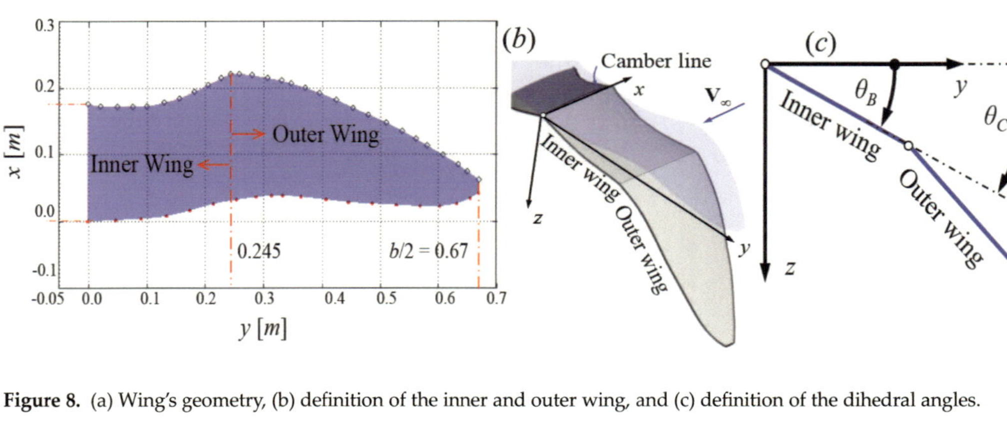

Figure 8. (a) Wing's geometry, (b) definition of the inner and outer wing, and (c) definition of the dihedral angles.

Case	θ_B^S	θ_C^S	T_I	T_S	T_F	V_8	α
I	−30°	30°	35 c_0/V_8	70 c_0/V_8	90 c_0/V_8	12.4 [m/s]	5°
II	10°	30°					
III	−10°	−30°					

Table 1. Simulation parameters.

In **Figure 9**, the lift coefficients for the three cases are plotted as functions of time. There are only slight differences between the solutions obtained with conditions (i) and (ii); thus, the vorticity being shed from the wingtips may be ignored. For all three cases, the number of elements in the wake associated with the wingtips is 20% of the total in the wake; when the wing-tip elements are eliminated the computational time is reduced by 35%.

During the first part of the flight simulation, when the wing is fully extended, the lift rapidly increases until the steady state is reached at T_I. During the second part of the flight $T_I < t < T_S$, which is the morphing process, the lift depends on the prescribed kinematics of the dihedral angles. In the third phase of the process after the morphing is completed, $t > T_S$, the wings approach new steady states. In Case III, the lift smoothly decreases until it reaches a local minimum, $t \approx T_M = (T_I + T_s)/2$, and then it increases to a value that corresponds to the wing's configuration in the new steady state. In this case, it appears that during the morphing process there could be a loss in altitude or perhaps a decrease in the rate of climb or an increase in the rate of descent. In Case II, the lift increases during the morphing process to reach a local maximum, $t \approx T_M$, and then it begins to decrease until it reaches a value that corresponds to the wing's new configuration in the steady state. In this case, it appears that the morphing process would produce a gain in altitude. In Case I, the lift behavior is different; it decreases smoothly until the steady-state solution is reached. For this last case, the lift characteristic does not have either a local maximum or

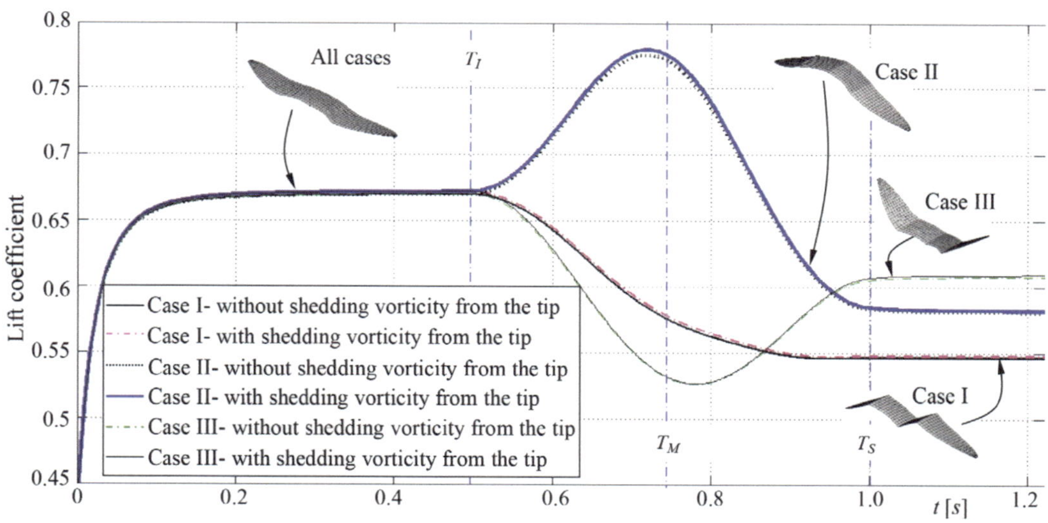

Figure 9. Lift coefficient as a function of time.

a local minimum during the morphing. With some tweaking in the three cases, the wing could arrive at any (within reason) prescribed steady-state lift, but most likely with the aircraft at different altitudes and different airspeeds. Here, although the airspeed was held constant in these three examples, it could be changed step by step. The present model provides an accurate tool to account for the following: (i) the strong dependence between the unsteady aerodynamic loads and the morphing motion, and (ii) the power needed to drive the morphing [61].

6.2.2. Aeroelasticity of morphing wings

The wing model adopted for the numerical aeroelastic simulations presented in this section consists of an inner and outer wing as it was described in Section 6.2.1. The wing's structural skeleton consists of a set of structural elements modeled as beams with rectangular cross sections (width = 2 cm and height = 0.6 cm) and material properties: density $\rho_b = 1187$ kg/m^3, Young's modulus $E = 3.18$ GPa, and shear modulus $G = 1.35$ GPa (see **Figure 10**).

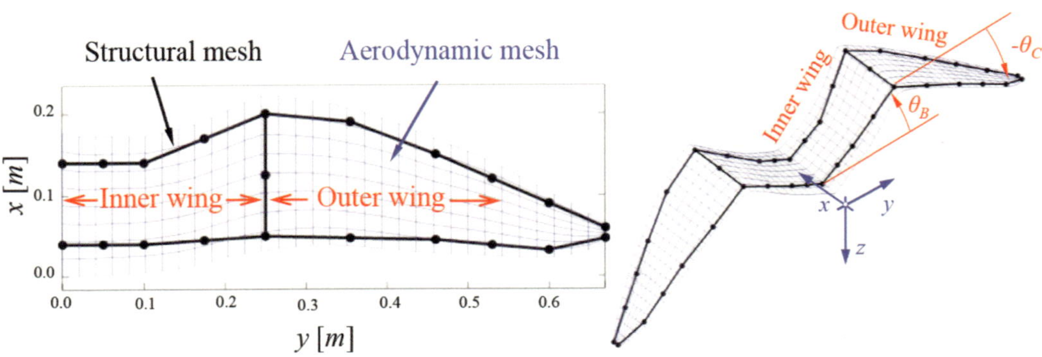

Figure 10. Computational wing model.

The inner and outer wing sections are oriented in space through the dihedral angles θ_B and θ_C. However, the wings do not have a prescribed motion (pitching, flapping, and lagging) and the wing's roots are considered to be clamped to the fuselage. This dynamic configuration deactivates the large rotations and displacements associated with primary motions; therefore, the set of governing equations (12) is reduced to the well-known set of classical ODEs for structural dynamics,

$$\mathbf{m}\,\ddot{\mathbf{p}} + \mathbf{K}\,\mathbf{p} = \left(Q_p\right)^T. \tag{20}$$

After algebraic manipulations and after expressing secondary motions as a linear combination of the free-vibration modes of the structure, the set of Eq. (20) is rewritten in a dimensionless form

$$d_{\hat{t}\hat{t}}\,\hat{\mathbf{p}}(\hat{t}) + \hat{\Lambda}\,\hat{\mathbf{p}}(\hat{t}) = \left(\frac{1}{2}\rho_C\,L_C^4\right)\tilde{\mathbf{m}}^{-1}\,\mathbf{\Psi}^T\left(\mathbf{G}_{AS}^{CP}\right)^T\hat{\mathbf{F}}_A, \tag{21}$$

where $\hat{t}$ is the Dimensionless-time (t/T_C); $d_{\hat{t}\hat{t}}$ denotes the second dimensionless time derivative; $\hat{\mathbf{p}}$ is a column vector containing the modal coordinates associated with the free vibration modes; $\tilde{\mathbf{m}}$ is the diagonal modal mass matrix; $\hat{\mathbf{\Lambda}}$ is a diagonal matrix that contains the natural frequencies in a dimensionless form (reduced frequency); $\mathbf{\Psi}$ is the modal matrix; L_C, T_C and ρ_C are characteristic properties of the problem [35]; and $\hat{\mathbf{F}}_A$ is the dimensionless form of the aerodynamic loads [49]. The rest of the variables are as defined before.

The main goal of the numerical experiment is to compute the flutter velocity V_F for different dihedral angles, ranging from $\theta_B = 0°$ to $\theta_B = 45°$, with steps of $\Delta\theta_B = 5°$. $\theta_C = 0°$ was used for all values of θ_B. Due to the lack of primary motions, the dihedral angles remain constant during the whole simulation. Additionally, the angle of attack is $\alpha = 0°$ for all simulations. The computational model consists of the following: (i) a finite-element mesh for the structural discretization consisting of 21 beam elements and (ii) a vortex-lattice mesh for the aerodynamic surface consisting of 256 panels (8 in the chordwise direction and 32 in the spanwise directions). In **Figure 11**, the authors show the flutter speed versus the dihedral angle of the inner wing θ_B. It can be observed that V_F decreases linearly from the maximum value of 29.66 m/s (for $\theta_B = 0°$), between 10° and 30°, to the minimum of 18 m/s for $\theta_B = 45°$. The curve represents a boundary for the stable and unstable regions. All flight conditions located below the curve lead to a response that tends to a static equilibrium position for any perturbation. On the contrary, the response for flight conditions located above the curve achieves a periodic motion with an amplitude that depends on the flight speed, that could lead to a collapse of the structure.

In **Figure 12a**, the amplitudes of three vibration modes are shown as functions of dimensionless time for a flight condition (pointed out in **Figure 11** by **a**) located in the stable region

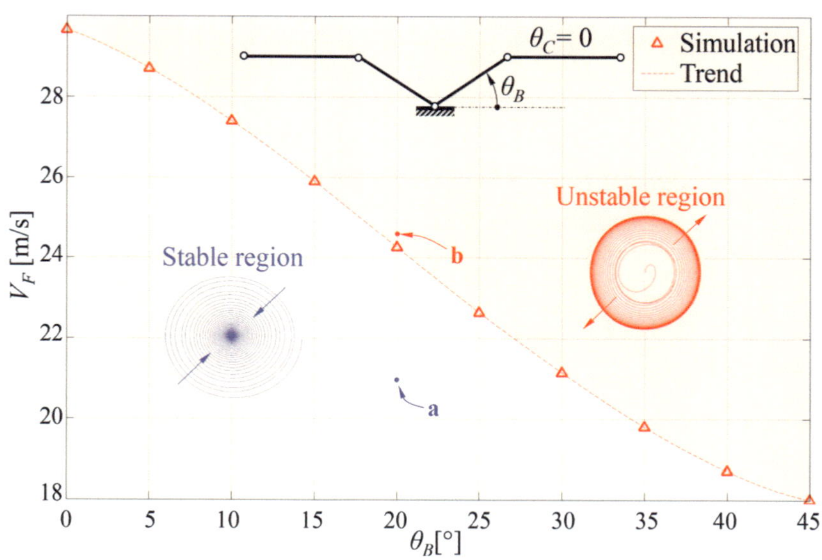

Figure 11. Flutter speed versus dihedral angle of the inner wing.

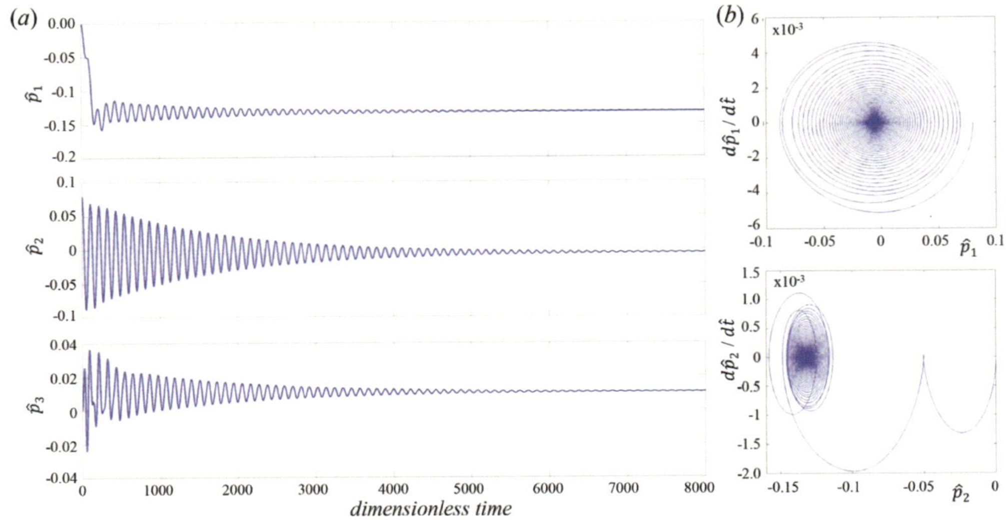

Figure 12. (a) Aeroelastic response for three modes as function of the dimensionless time. (b) Phase portrait for two modes ($V_\infty = 21.5$ m/s and $\theta_B = 20°$).

(free stream of 21 m/s and a configuration defined by $\theta_B = 20°$). An initial perturbation on the second mode $\hat{p}_2(0) = 0.08$ was imposed in the simulation. The amplitudes of all vibration modes clearly decay, not to zero, but to their equilibrium positions. This is exclusively due to the aerodynamic damping, because the structural model does not include any damping. In

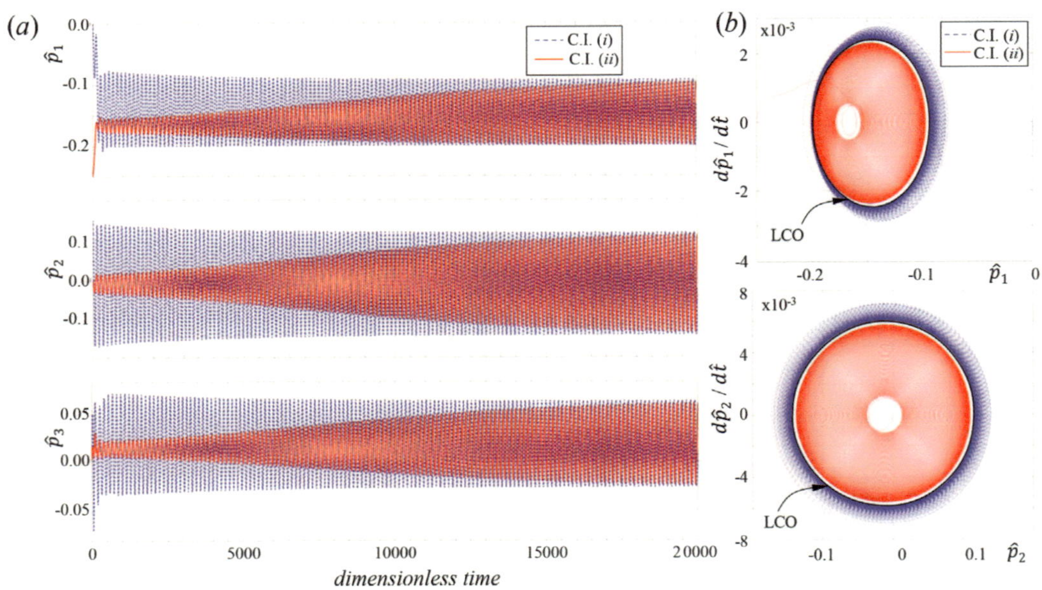

Figure 13. (a) Aeroelastic response for three modes as function of dimensionless time. (b) Phase portrait for two modes ($V_\infty = 24.5$ m/s and $\theta_B = 20°$ for both responses, blue-dashed line and red-solid line).

Figure 12b, the authors show the phase portraits for two modes. Both modes have a behavior similar to a stable focus.

In **Figure 13a**, the amplitudes of the same three modes are also shown as functions of time but for a flight condition located in the unstable region (pointed out in **Figure 11** by **b**, with a free stream of 24.5 m/s and a configuration defined by $\theta_B = 20°$). Two initial conditions were considered in the simulation. The corresponding responses (blue-dashed line and red-solid line) are characterized by a transient behavior that evolves into periodic orbits whose amplitudes are the same. In **Figure 13b**, the phase portraits for two of the modes clearly reveal the presence of a limit-cycle oscillation (LCO). The trajectory associated with the first initial condition approaches the LCO from outside, while the trajectory associated with the other initial condition approaches the LCO from inside.

7. Concluding remarks

A combined structural and aerodynamic model for studying the aeroelastic behavior of flying robots with flexible wings has been presented, and aeroelasticity studies have been carried out for systems with flexible wings. The modular approach is based on the following: (i) a general, nonlinear-unsteady vortex-lattice method, (ii) a segregated version of Lagrange's equations and the Floating Frame of Reference formalism for constrained systems, (iii) tight coupling of the aerodynamic model (UVLM) with the structural model, (iv) radial basis functions to transfer information between the aerodynamic and structural grids, and (v) an efficient and novel algorithm, based on a computational platform with modular structure, to integrate the equations interactively and simultaneously in the time domain.

As illustrated by the examples, the authors' methodology will enable fully coupled aeroelastic simulations for MAVs with flapping wings and/or morphing wings and further investigations in this direction. Their development paves the way to advance the current methodology by implementing a vortex-particle method to improve the description of the wakes and a fast multi-pole method for computing the velocity field, and by including various other beam and shell elements.

Acknowledgements

The authors gratefully acknowledge the partial support received from the Consejo Nacional de Investigaciones Científicas y Técnicas, Argentina, the U.S. National Science Foundation through Grant No. CMMI-1250187, the U.S. Air Force Office of Scientific Research through Grant No. FA95501510134, and the Minta Martin Foundation. In addition, the authors would like to thank the Grupo de Electrónica Aplicada (GEA) and Grupo de Matemática Aplicada (GMA), Engineering School, Universidad Nacional de Río Cuarto, Argentina.

Author details

Sergio Preidikman[1,2*], Bruno Antonio Roccia[1,2,3], Marcos Leonardo Verstraete[2,3], Marcelo Federico Valdez[2,4], Dean T. Mook[5] and Balakumar Balachandran[6]

*Address all correspondence to: spreidik@umd.edu

1 Facultad de Ciencias Exactas, Físicas y Naturales, Universidad Nacional de Córdoba, Argentina

2 Consejo Nacional de Investigaciones Científicas y Técnicas, CONICET, Argentina

3 Grupo de Matemática Aplicada, Facultad de Ingeniería, Universidad Nacional de Río Cuarto, Argentina

4 Universidad Nacional del Salta, Argentina

5 Department of Biomedical Engineering and Mechanics, Virginia Polytechnic Institute and State University, Blacksburg, VA, USA

6 Department of Mechanical Engineering, University of Maryland, College Park, MD, USA

References

[1] Dickinson MH, Lehmann FO, Sane SP. Wing rotation and the aerodynamic basis of insect flight. Science. 1999;284:1954–1960

[2] Ellington CP, Van Den Berg C, Willmott AP, Thomas ALR. Leading-edge vortices in insect flight. Nature. 1996;384:626–630

[3] Wang ZJ, Birch JM, Dickinson MH. Unsteady forces and flows in flow Reynolds number hovering flight: Two-dimensional computational vs robotic wing experiments. Journal of Experimental Biology. 2004;207:269–283.

[4] Weis-Fogh T. Quick estimates of flight fitness in hovering animals, including novel mechanisms for lift production. Journal of Experimental Biology. 1973;59:169–230

[5] Ellington CP. The aerodynamics of hovering insects flight. III. Kinematics. Philosophical Transactions of the Royal Society of London. Series B, Biological Sciences. 1984;305(1122):41–78

[6] Willmott AP, Ellington CP. The mechanics of flight in the hawkmoth Manduca sexta. I. Kinematics of hovering and forward flight. Journal of Experimental Biology. 1997;200:2727–2738

[7] Wang H, Zeng L, Liu H, Yin C. Measuring wing kinematics, flight trajectory and body attitude during forward flight and turning maneuvers in dragonflies. Journal of Experimental Biology. 2003;206:745–757

[8] Walker SM, Thomas ALR, Taylor GK. Deformable wing kinematics in free-flying hoverflies. Journal of the Royal Society of London, Interface. 2009b;7(42):131–142

[9] Lehmann FO, Gorb S, Nasir N, Schützner P. Elastic deformation and energy loss of flapping fly wings. Journal of Experimental Biology. 2011;214:2949–2961

[10] Mountcastle AM, Combes SA. Wing flexibility enhances load-lifting capacity in bumblebees. Proceedings of the Royal Society, Series B, Biological Sciences. 2013;280(1759):1–8

[11] Ishilhara D, Hoire T, Denda M. A two-dimensional computational study on the fluid-structure interaction cause of wing pitch changes in dipteran flapping flight. Journal of Experimental Biology. 2009;212:1–10

[12] Vanella M, Fitzgerald T, Preidikman S, Balaras E, Balachandran B. Influence of flexibility on the aerodynamic performance of a hovering wing. Journal of Experimental Biology. 2009;212:95–105

[13] Fitzgerald T, Valdez M, Vanella M, Balaras E, Balachandran B. Flexible flapping systems: Computational investigations into fluid-structure interactions. The Aeronautical Journal. 2011;115(1172):593–604

[14] Smith MJC. The effect of the flexibility on the aerodynamics of moth wing: Towards the development of flapping-wing technology. In: 33rd Aerospace Sciences Meeting and Exhibit; 9–12 January; Reno: AIAA 95-0743; 1995

[15] Willis DJ, Israeli ER, Persson P, Drela M, Peraire J, Swartz SM, Breuer KM. A computational framework for fluid structure interaction in biologically inspired flapping flight. In: 25th AIAA Applied Aerodynamics Conference; 25–28 June; Miami: AIAA 2007-3803; 2007.

[16] Kim DK, Lee JS, Lee JY, Han JH. An aeroelastic analysis of a flexible flapping wing using modified strip theory. In: SPIE 15th Annual Symposium Smart Structures and Materials; 09 March; San Diego, California: 69281O; 2008.

[17] Gopalakrishnan P, Tafti DK. Effect of wing flexibility on lift and thrust production in flapping flight. Journal of Aircraft. 2010;48(5):2505–2519

[18] Nakata T, Liu H. A fluid-structure interaction model of insect flight with flexible wings. Journal of Computational Physics. 2012;231:1822–1847

[19] Liu H. Integrated modeling of insect flight: From morphology, kinematics to aerodynamics. Journal of Computational Physics. 2009;228:439–459

[20] Chimakurthi SK, Stanford BK, Cesnik CES, Shyy W. Flapping wing CFD/CSD aeroelastic formulation based on a co-rotational shell finite element. In: 50th AIAA/ASME/ASCE/

AHS/ASC Structures, Structural Dynamics, and Materials Conference, AIAA Paper 2009–2412; 4–7 May 2009; Palm Springs, California

[21] Shyy W, Udaykumar H, Rao M, Smith R. Computational Fluid Dynamics with Moving Boundaries. New York, NY: Dover; 2007

[22] Malhan R, Baeder JD, Chopra I, Masarati P. CFD-CSD coupled aeroelastic analysis of flexible flapping wings for MAVs applications. In: 54th AIAA/ASME/ASCE/AHS/ASC Structures, Structural Dynamics, and Materials Conference; 8–11 April 2013; Boston

[23] Lakshminarayan VK, Baeder JD. Computational investigation of micro-scale coaxial rotor aerodynamics in hover. Journal of Aircraft. 2010;47(3):940–955

[24] Masarati P, Morandini M, Quaranta G, Vescovini R. Multibody analysis of a micro-aerial vehicle flapping wing. In: Multibody Dynamics 2011; July 2011; Brussels

[25] Unger R, Haupt MC, Horst P, Radespiel R. Fluid-structure analysis of a flexible flapping airfoil at low Reynolds number flow. Journal of Fluid and Structures. 2012;28:72–88

[26] Kroll N, Rossow CC, Schwamborn D, Becker K, Heller G. MEGAFLOW–a numerical flow simulation tool for transport aircraft design. In: 23rd International Congress of Aeronautical Sciences ICAS. 8–13 September; Toronto, Canada: Paper 2002–1105; 2002.

[27] ANSYS Inc. ANSYS11.0documentation. 2006. Available from: http://www.ansys.com

[28] Bose C, Badrinath S, Gupta S, Sarkar S. Dynamical stability analysis of a fluid structure interaction system using a high fidelity Navier-Stokes solver. Procedia Engineering. 2016;144:883–890

[29] OpenFOAM. The Open Source CFD Toolbox User Guide. 2013. Available from: URL: http://www.openfoam.org/

[30] Nayfeh AH, Balachandran B. Applied Nonlinear Dynamics: Analytical, Computational, and Experimental Methods. New York, NY: Wiley; 1995 (2006)

[31] Murua J, Palacios R, Graham JMR. Applications of the unsteady vortex-lattice method in aircraft aeroelasticity and flight dynamics. Progress in Aerospace Sciences. 2012;55:46–72

[32] Preidikman S, Mook DT. Time–Domain simulations of linear and non-linear aeroelastic behavior. Journal of Vibration and Control. 2000;6(8):1135–1176

[33] Obradovic B, Subbarao K. Modeling of flight dynamics of morphing-wing aircraft. Journal of Aircraft. 2011;48(2):391–402

[34] Taha HE, Hajj MR, Nayfeh AH. Flight dynamics and control of flapping-wing MAVs: A review. Journal of Nonlinear Dynamics. 2012;7(2):907–939

[35] Preidikman S. Numerical simulations of interactions among aerodynamics, structural dynamics, and control systems [Ph.D. thesis]. Blacksburg: Virginia Polytechnic Institute and State University; 1998

[36] Kalmar-Nagy T, Stanciulescu I. Can complex systems really be simulated? Applied Mathematics and Computation. 2014;227:199–211

[37] Roccia BA, Preidikman S, Balachandran B. Computational dynamics of flapping wings in hover flight: A co-simulation strategy. AIAA Journal. In Press, pp. 1–17, 2017.

[38] Roccia BA, Preidikman S, Massa JC, Mook DT. A modified unsteady vortex-lattice method to study the aerodynamics of flapping-winds in hover flight. AIAA Journal. 2013;51(11):2628–2642

[39] Konstandinopoulos P, Mook DT, Nayfeh AH. A numerical method for general, unsteady aerodynamics. In: 7th Atmospheric Flight Mechanics Conference; 19–21 August; Albuquerque, New Mexico: AIAA-81–1877; 1981

[40] Van Garrel A. The Development of a Wind Turbine Aerodynamics Simulation Module, ECN Rept. ECN-C-03-079, Delft Univ. of Technology, Delft; 2003

[41] Katz J, Plotkin A. Low-Speed Aerodynamics. 2nd ed. New York, NY: Cambridge University Press; 2001. pp. 421–495

[42] Mook DT, Maddox SA. Extension of a vortex-lattice method to include the effects of leading-edge separation. Journal of Aircraft. 1974;11(2):127–128

[43] Shabana AA. Dynamics of Multibody Systems. 3rd ed. Cambridge: Cambridge University Press; 2010

[44] Bauchau OA. Flexible Multibody Dynamics. New York, NY: Springer; 2011

[45] Cook RD, Malkus DS, Plesha ME, Witt RJ. Concepts and Applications of Finite Element Analysis. 4th ed. New York, NY: Wiley; 2001.

[46] Beckert A, Wendland H. Multivariate interpolation for fluid-structure-interaction problems using radial basis functions. Aerospace Science and Technology. 2001;5: 125–134

[47] Buhmann M. Radial Basis Functions. Cambridge: Cambridge University Press; 2005

[48] Wendland H. Scattered Data Approximation. Cambridge: Cambridge University Press; 2005

[49] Verstraete ML. Simulaciones numéricas del comportamiento aeroelástico de vehículos aéreos no tripulados con alas que cambian de forma [Ph.D. Dissertation]. Argentina: Engineering School, National University of Rio Cuarto; 2016

[50] Baumgarte J. Stabilization of constraints and integrals of motion in dynamical systems. Computational Mathematics Applied to Mechanical Engineering. 1972;1:1–16

[51] Ascher UM, Chin H, Petzold LR, Reich S. Stabilization of constrained mechanical systems with DAEs and invariant manifolds. Journal of Mechanics of Structures and Machines. 1995;23:135–158

[52] Carnahan B, Luther HA, Wilkes JO. Applied Numerical Methods. New York, NY: John Wiley and Sons; 1969

[53] Roccia BA, Preidikman S, Massa JC, Mook DT. Development of a kinematical model to study the aerodynamics of flapping-wings. International Journal of Micro Air Vehicles. 2011;3(2):61–88

[54] Sun M, Tang J. Unsteady aerodynamic force generation by a model fruit fly wing in flapping motion. Journal of Experimental Biology. 2002;205:55–70

[55] Ansari SA, Żbikowski R, Knowles K. Non-linear unsteady aerodynamics model for insect-like flapping wings in the hover. Part 1: Methodology and analysis. Journal of Aerospace Engineering. 2006;220:61–83

[56] Ansari SA, Żbikowski R, Knowles K. Non-linear unsteady aerodynamics model for insect-like flapping wings in the hover. Part 2: Implementation and validation. Journal of Aerospace Engineering. 2006;220:169–186

[57] Roccia BA, Preidikman S, Verstraete ML, Mook DT. Influence of spanwise twisting and bending on lift generation in MAV-like flapping wings. Journal of Aerospace Engineering (ASCE), paper 04016079; 2016. pp. 1–17

[58] Fry SN, Sayaman R, Dickinson MH. The aerodynamics of hovering flight in Drosophila. Journal of Experimental Biology. 2005;208:2303–2318

[59] Tennekes H. The Simple Science of Flight: From Insects to Jumbo Jets. Cambridge: MIT Press; 2009

[60] Liu T, Kuykendoll K, Rhew R, Jones S. Avian wing geometry and kinematics. AIAA Journal. 2006;44(5):954–963

[61] Verstraete ML, Preidikman S, Roccia BA, Mook DT. A numerical model to study the nonlinear and unsteady aerodynamics of bioinspired morphing-wing concepts. International Journal of Micro Air Vehicles. 2015;7(3):327–345

Chapter 2

Overview of Coandă MAV as an Aerial Robotic Platform

Harijono Djojodihardjo

Additional information is available at the end of the chapter

http://dx.doi.org/10.5772/intechopen.70157

Abstract

With the increasing need of micro-air-vehicles (MAVs) and advances in MAV technology, Coandă MAVs offer new promises and challenges. In this context, Coandă MAVs capabilities are analyzed. As a baseline, a mathematical model for a spherical Coandă MAV in hover and translatory motion is developed and analyzed from first physical principles. A computational fluid dynamic (CFD) simulations for a Coandă MAV generic model are carried out to assess the theoretical prediction and obtaining further physical insight on the Coandă MAV flow physics. The mathematical model and performance measures are developed to assess the capability of the semi-spherical Coandă MAV in performing effective flight as an aerial robotic platform, as indicated by the relationships between the relevant parameters of the mathematical model of the Coandă MAV to its system of flight forces.

Keywords: aerial robot, Coandă micro air vehicle (CMAV), Coandă MAV aerodynamics, Coandă MAV flight dynamics, unmanned micro air vehicle (UMAV)

Subject Area: Aerodynamics, Aerospace Engineering, Applied Mathematics and Physics, Flight Mechanics

1. Introduction

The utilization of Coandă effect and Coandă jet [1, 2] for flight vehicles have been carried out for a range of aircrafts and vehicles for decades and for future aircrafts [3–14], due to its flow control effectiveness. Other applications are potential in wind-turbines technology [15–20], and the Coanda effect have also been found and utilized in many other engineering applications and health science, such as in the upper respiratory system [20–24]. The science of flow control can be considered to originate from Prandtl [25] seminal paper in 1904, which has been hailed as a breakthrough in the science of fluid mechanics by the introduction of the boundary-layer theory. Prandtl's theory can be accounted for the physics of the separation phenomena and the boundary layers control.

INTECH open science | open minds

© 2017 The Author(s). Licensee InTech. This chapter is distributed under the terms of the Creative Commons Attribution License (http://creativecommons.org/licenses/by/3.0), which permits unrestricted use, distribution, and reproduction in any medium, provided the original work is properly cited. (cc) BY

The Coandă effect is based on the property of a jet flow to attach itself to a curved surface as it tangentially blows along and to remain attached until it leaves the surface or reach a critical stability breakdown. Coandă jet is one of the flow control techniques for modifying and enhancing aerodynamic performance, stability and control. Coandă jet has also been applied in the development of novel aircrafts for short take-off and landing. With the progress of many enabling technologies and need for disaster mitigation, security, and environmental conservation, the relatively new application of Coandă effect in micro-air-vehicles (MAVs) as robotic platforms is very challenging. Capitalizing on the basic fundamental principles, the aero- and flight dynamics of a baseline Coandă MAV configuration are analyzed and developed to formulate the relationships between various essential parameters to the equation of motion in hover, and translatory motion, which should give further insight on and can be further elaborated into other Coandă MAV characteristics and performance related to maneuver and stability and control.

Coandă micro-air-vehicles have been developed in various configurations. To meet the desired Coandă MAV operation, mission, and design requirements, one should establish the basic working relationships among various significant design parameters and variables to the generated aerodynamics forces. For this purpose, several approaches can be followed. The most logical first approach is the analytical tool based on basic fundamental principles. With the impressive progress of computational fluid dynamics (CFD) which has brought with it flow visualization capabilities; CFD utilization will assist the identification to the problem and assessment of the analytical approach. Furthermore, CFD visualization facilitates better insight and identification of salient details. Another important approach is the use of experimental tools, which is the foundation of scientific observation through meticulous experimental design and has intricate relationships with analytical approach. For the conceptual and prototype design stages, analytical, and CFD computational and visualization tools will expedite the efforts to that end.

The theoretical analysis, mathematical modeling and synthesis or Coandă micro-air-vehicles as an aerial robot platform will start with the identification and definition of a Coandă jet as a relatively thin and slightly viscous jet flowing over and adhere on a smooth curved surface. The main physical parameters are the angle of separation, θ; slot width, b; radius of curvature, R; Reynolds number, Re; the Coandă jet velocity, pressure differential $p_s - p_\infty$, Reynolds number and the Coandă jet momentum coefficient. Various platforms will be discussed to define the merit of Coandă MAV as a robotic platform. Hovering semi-spherical Coandă MAV will be utilized in the baseline aerodynamic analysis. The mathematical model of Coandă MAV to generate the equation of motion starts with first principles and the basic flight dynamic equation. CFD simulation and demonstration experiment are elaborated to assess the functionality of Coandă MAV aerial robot platform.

In the model development and analysis of a semi-spherical Coandă MAV, a propulsion system is utilized to introduce a Coandă jet blanket which is deflected downward due to the curved upper surface of the vehicle, to provide both hover as well as cruise capabilities. Such Coandă jet has been utilized for circulation enhancement for fixed-wing aircraft in forward flight and turbine blades movement [4–9, 12–20]. Thus, lift will be generated [27–35], which can be utilized for hover and later for cruise propulsion [10, 35, 36]. The principle of Coandă MAV lift generation as well as the equation of motion for its translatory motion will be derived and elaborated.

Some results from the authors' previous analysis [31–34] provide the establishment of Coandă MAV spherical configuration baseline. The equations derived for the baseline configuration are significant for establishing preliminary design configuration concepts, and can be readily modified for other configurations for assessment, comparative and improvement purposes. CFD visualization studies will be instrumental in gaining in-depth understanding of the analytical and physical model development for further analysis.

Based on the physical and mathematical model conceived, some basic results were obtained to describe the physics of the flow field associated with the Coandă effect jet sheets. Having developed the basic fluid-dynamic analysis of the Coandă MAV lift generation, the equation of motion for Coandă MAV in translatory motion can be derived.

Various CFD studies have also indicated that there are still significant discrepancies between CFD and experimental studies [5, 15], which necessitate the existence of specific baseline configurations for validation purposes. Coandă MAV is expected to be capable of maneuvering as illustrated in **Figure 1 (a)** and **(b)**, and the analysis carried out here is associated with hovering condition, which should give further perception into its hovering performance.

The analysis carried out here is associated with hovering and cruising conditions, which should give further awareness on and can be further elaborated into Coandă MAV maneuvering performance. The baseline analysis on Coandă MAV spherical configuration carried out here refers to and confirms the authors' previous analysis [31–35].

The generic and baseline physical and mathematical model developed have been utilized to obtain some basic results that are essential in describing the flow field and the physical phenomena related to the relevant surfaces influenced by the Coandă effect jet sheets. Comparison of the numerical computation results for some baseline cases with experimental data under similar conditions will be essential, while computational parametric study will be

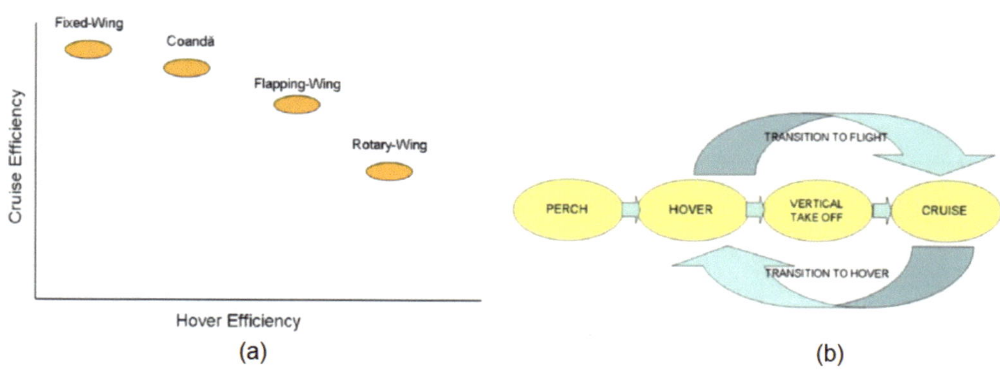

Figure 1. (a) An impression of the possible Coandă MAV qualitative performance in comparison to other flight vehicles (adapted from [10]), (b) Flight manoeuvring structure (adapted from [26]).

helpful as a preliminary effort for design and performance optimization. Some examples of Coandă effect applications and developments are illustrated in **Figure 2**.

It is with such objectives in mind that parametric studies may offer some clues on relevant parameters which may be utilized in a multivariable optimization (and to a larger scale, multidisciplinary optimization). The introduction of Coandă jet on both airfoils and aerodynamic surfaces results in enhanced L/D, which depends on the jet velocity or momentum coefficient. TE (Trailing-edge) rounding-off of in the introduction of the Coandă jet has been shown to be effective in increasing airfoil, L/D, as exemplified by recent applications of Coandă jet for STOL (Short Take-Off and Landing) or ESTOL (Extreme Short Take-Off and

	Some Coandă Effect Development and Applications		Examples/ Remarks
1	Some Coandă Inventions	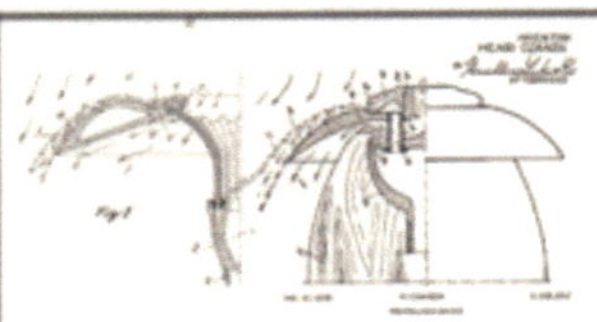	
2	Willard Custer, Englar and Campbell, Coandă-Effect Aircraft		Custer Channel Wing Aircraft; Lift due to " the speed of air, not the airspeed"
3	John Carver Meadow Frost Coandă-Effect Aircraft		Avrocar
4	Enhanced Lift by Coandă-Effect Circulation Enhancement		Antonov An-72
5			Boeing YC-14
6	Co-Flow Jet		US Navy Coandă-Effect Aircraft
7	California Polytechnique Group (Marshall, de la Montanya, etc)		Combined Blowing Circulation Control Application yo Extreme Short Take-Off & Landing
8	Automotive Application		Increase of base pressure, Coandă-Effect through exhaust gas for better traction

Figure 2. Some examples of Coandă effect applications and development (adapted from [14]).

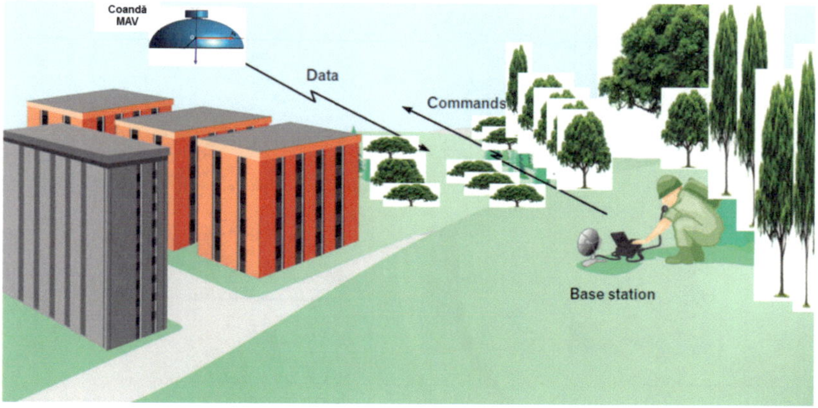

Figure 3. Micro-air-vehicle (MAV) used for reconnaissance. The MAV and a portable base station could remotely monitor the MAV under hostile conditions. Sensor data could be transferred in real time or stored on board the MAV (adapted from [37]).

Landing), e.g., S809. The study on Coandă jet application to wind turbines that could produce maximum total energy output of Coandă configured airfoil exceeding that predicted by Betz limit could provide further insight on Coandă jet application for MAV.

The test and evaluation of unmanned systems presents special challenges, but these challenges are amplified when one moves into the realm of micro-air-vehicles. Further complications arise when these MAVs are fully autonomous [37]. Autonomous MAVs must be able to sense the environment and guide their movement through it. To find their targets, autonomous MAVs should be able to identify them using object recognition procedure, and move toward their goal based on a path planning algorithm. The associated sensors can be employed for an aerial robotic MAV. MAVs have been utilized for surveillance that uses a data link and line-of-sight control, as illustrated in **Figure 3** (adapted from Davis et al. [37]); advanced MAVs can hover and navigate independently and carry multiple sensors. In addition, the production in large quantities of MAVs at low unit cost can be facilitated by exploiting microfabrication technology.

2. Coandă MAV Aerodynamics of hovering semi-spherical Coandă MAV as a baseline configuration

The Coandă MAV mathematical model utilized here follows closely that of Djojodihardjo et al. [31, 32] and Ahmed et al [33, 34]. The model was conceived to be generic and basic, and allow the application of first principles. The detail is summarized and reproduced here as a baseline reference for developments that may follow. In the Coandă MAV configuration illustrated in **Figure 4**, an actuator rotor is located at the center of the upper part of the body. As a baseline, the dimension of the rotor can be designed to be small as required. As illustrated in the figure, the flow being drawn by the actuator is entirely utilized for lift, by producing radial flow on the surface of the body as a Coandă jet blanket. In addition, like in a helicopter, a part of the flow produced by the actuator can pass through a vertical stream tube to produce lift (or thrust).

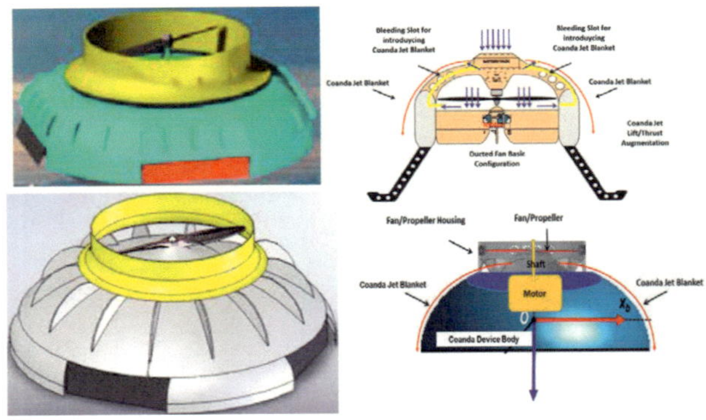

Figure 4. Various semi-spherical Coandă Devices utilizing a propeller to develop Coandă jet over its upper surface.

In the analysis of Coandă effect, momentum conservation principle in the sense of Euler equation is applied in the analysis to find the relevant aerodynamic forces and to define the performance parameters of the Coandă MAV. This procedure, applied for a baseline and simplified semi-spherical Coandă MAV configuration, is shown in **Figure 5**.

A rigorous analysis is then carried out to see how Coandă effect contributes to the Coandă MAV lift generation capability. For this purpose, the fundamental conservation analysis is applied on the control volume CV (dashed rectangle in **Figure 5**).

Applying the continuity equation (that is the mass conservation equation) given by $\dot{m}_{J.in} = \dot{m}_{J.out} = \dot{m}$, and assuming incompressible flow throughout, on the control volume enclosing the Coandă Blanket as depicted in **Figure 6**, there is obtained

$$\rho \cdot V_{j\text{-}R} \cdot 2\pi Rh = \rho \cdot V_{j\text{-}in} \cdot 2\pi R_i h_i = \rho \cdot V_{j\text{-}out} \cdot 2\pi R_o h_o \tag{1}$$

and

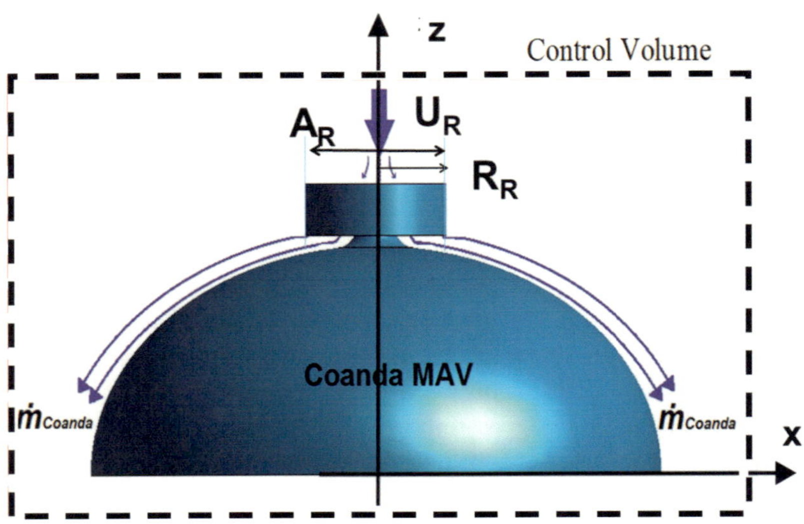

Figure 5. An equivalent semi-spherical Coandă MAV for Generic Analysis.

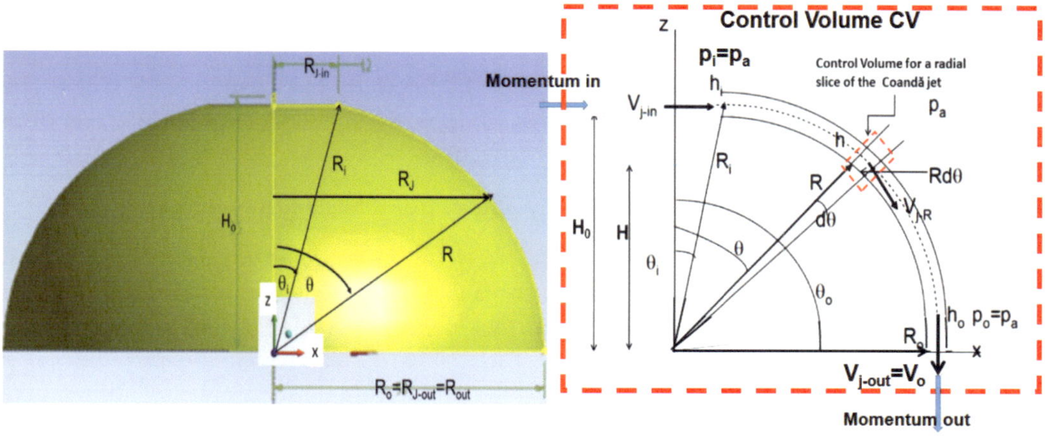

Figure 6. Schematic of Coandă jet blanket for a spherical MAV momentum balance.

$$\begin{aligned} V_{j\text{-}R} &= \frac{R_i h_i}{R h_R} V_{j\text{-}in} = \frac{\dot{m}}{2\pi\rho R h_R} \\ V_{j\text{-}out} &= \frac{R_i h_i}{R_o h_o} V_{j\text{-}in} = \frac{\dot{m}}{2\pi\rho R_o h_o} \end{aligned} \tag{2}$$

where $V_{j\text{-}R}$ is the jet flow velocity, R is the vehicle body radius, h is the jet slot thickness and $\dot{m}$ is the jet mass flow rate. $V_{J\text{-}out}$, R_o, h_o, $V_{J\text{-}in}$ R_i, and h_i refer to velocity, radius and jet thickness at the outlet and inlet, respectively, of the Coandă jet blanket. As depicted in **Figure 6**, the, the Coandă jet blanket is exposed to the ambient pressure pa, which is then the prevailing static pressure on it. The momentum equation applied to the control volume in the z (vertical) direction can be differentiated into the contribution due to the Coandă Blanket momentum and the pressure difference on the body due to Coandă Blanket:

$$\begin{aligned} \begin{pmatrix} \text{Total Lift force} \\ \text{due to Coandă Blanket} \end{pmatrix} &= \begin{pmatrix} \text{Vertical component of momentum} \\ \text{flux balance due to Coandă Blanket} \end{pmatrix} \\ &\quad + \begin{pmatrix} \text{Pressure difference on the body} \\ \text{of MAVsubject to Coandă Blanket} \end{pmatrix} \end{aligned} \tag{3}$$

The contribution of lift from the momentum flux through the control volume CV in the y (vertical) direction is given by

$$\begin{pmatrix} \text{Force on the } \textit{Coandă MAV} \\ \text{in the z-direction} \end{pmatrix} = \begin{pmatrix} \text{Rate of Momentum In} \\ \text{into CV the y-direction} \end{pmatrix} - \begin{pmatrix} \text{Rate of Momentum Out} \\ \text{of CV in the z-direction} \end{pmatrix} \tag{4}$$

with z and force positive in the upward direction.

It is noted that the momentum in the radial direction does not contribute to lift. Therefore, as exhibited in **Figure 6**, the momentum equation in the y direction for the control volume CV is

$$F_{Coanda\,jet\,Blanket} = \dot{m}\,V_{J\text{-}out} = 2\pi R_i h_i\,\rho \cdot V_{J\text{-}in} \cdot V_{J\text{-}out} \tag{5}$$

Before proceeding to analyze the contribution of the forces on the Coandă MAV due to the pressure difference across the control volume, the energy equation applied to the control volume will be considered. It reads

Energy input to the control volume = Energy output from the control volume

In the control volume analysis throughout this work, uniform properties across the sectional areas at the input and output of the Coandă jet blanket have been assumed. The entrainment energy exchange between the ambient air and the Coandă jet blanket as well as other energy losses have been ignored.

Hence, referring again to **Figure 6**, taking into consideration the mass conservation and assuming similar ambient pressure at the inlet and outlet, the application of the energy conservation equation within the control volume which only involves the fluid flow along the Coandă jet blanket results in

$$\begin{gathered}
\frac{1}{2}m_{in}V_{J-in}{}^{2} = \frac{1}{2}m_{out}V_{J-out}{}^{2} \quad [M]\left[L^2T^{-2}\right] = \left[MLT^{-2}\right][L] \\
2\pi\rho V_{J-in}h_{J-in}R_{in}V_{J-in} = 2\pi\rho V_{J-out}h_{J-out}R_{out}V_{J-out} = 2\pi\rho V_J h_J R_J V_J \\
\dot{m}_{in}V_{J-in} = \dot{m}_{out}V_{J-out} = \dot{m}_J V_J{}^2 \\
V_{J-in} = V_{J-out} = V_J{}^2
\end{gathered} \tag{6}$$

Therefore, upon substitution of Eq. (6) into Eq. (5), one obtains

$$L_{\substack{Momentum \\ Coanda\,jet\,blanket}} = F_{Coanda\,jet\,Blanket} = \dot{m}\,V_{j\text{-}out} = 2\pi R_i h_i\,\rho \cdot V_{J\text{-}in}{}^2 \tag{7}$$

Using the model as shown in **Figure 7**, the underbody represented by the outer part of the lower surface of the control volume as well as the wake of the MAV are at ambient pressure. The entire flow entering the rotor represented by the actuator disk can be utilized for the Coandă jet blanket after leaving the actuator disk with velocity $V_{J\text{-}in}$, and then leaves the Coandă MAV after being ejected as $V_{J\text{-}out}$ and combined with the entrained fluid outside the Coandă jet blanket as the wake. If part of the flow is used to feed the Coandă jet blanket, then the rest of the flow leaving the actuator will form a stream tube leaving the Coandă MAV like a ducted fan and eventually will be combined with the exiting jet from the Coandă jet blanket.

Noting the specific characteristics of Coandă effect is attributed to the attachment of a jet to a curves surface, an analysis will be performed on the Coandă device control volume depicted in **Figure 6** by focusing on a control volume enclosing radial slice of the Coandă blanket with thickness h at (R, θ). Considering a balance of force along the radial direction R in the Coandă jet slice control volume, centrifugal effects along R allows a radial pressure gradient to prevail perpendicular to Coandă jet; hence

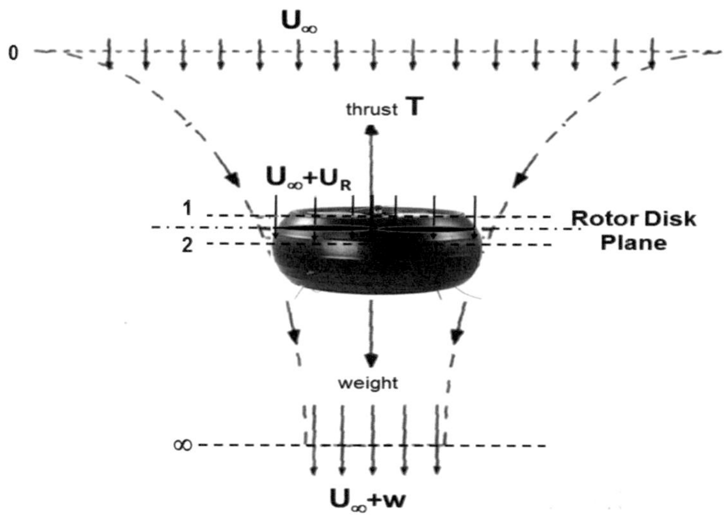

Figure 7. Control volume analysis of Coandă MAV actuator disk schematic.

$$\left(p_a - p_{\substack{J-\theta \\ Coandă-Surface}}\right) R_{J-\theta} d\theta = radial\text{-}acceleration \cdot \rho h_{J-\theta} R_{J-\theta} d\theta \tag{8a}$$

where

$$radial\text{-}acceleration = \frac{V_{J-\theta}{}^2}{R_{J-\theta}} \tag{8b}$$

so that

$$p_{\substack{J-\theta \\ Coandă\text{-}Surface}} = p_a - \frac{V_{J-\theta}{}^2}{R_{J-\theta}} \cdot \rho h_{J-\theta} \quad \left[\frac{MLT^{-2}}{L^2}\right] - \left[\frac{L^2T^{-2}}{L}\right]\left[\frac{ML}{L^3}\right] \tag{8c}$$

Here the ambient pressure p_a acts on the outer surface of the Coandă jet blanket, while $p_{J\text{-}\theta}$ acts on the lower surface of the Coandă jet blanket, thus on the upper surface of the Coandă MAV.

The contribution of the vertical forces on the Coandă MAV due to the pressure difference across the control volume can now be considered. This is then give by

$$L_{pressure} = Pressure\ Force_{CV-upper} - Pressure\ Force_{CV-lower} = 2\pi \int_{R_{J-\theta}\ at\ jat\ inlet}^{R_{J-\theta}=R_{J-out}} p_{J-\theta} R_{J-\theta} dR_{J-\theta} - p_a \pi R_{J-\theta}{}^2 \tag{9a}$$

Substituting Eq. (8c), the vertical force due to pressure difference is then given by

$$L_{pressure} = Force_{\substack{Pressure\\CV-upper}} - Force_{\substack{Pressure\\CV-lower}} = 2\pi\int_{R_{J-\theta}\ at\ jat\ inlet}^{R_{J-\theta}=R_{J-out}} \left(-\frac{V_{J-\theta}{}^2}{R_{J-\theta}}\cdot\rho h_{J-\theta}\right)R_{J-\theta}dR_{J-\theta} \tag{9b}$$

Since either the plenum chamber between the upper curved surface or the bottom part of the Coandă MAV is at the atmospheric pressure p_a, then p_a in Eq. (9b) will drop off. Hence

$$\begin{aligned} L_{pressure} = Force_{\substack{Pressure\\CV-upper}} - Force_{\substack{Pressure\\CV-lower}} &= 2\pi\int_{R_{J-\theta}\ at\ jat\ inlet}^{R_{J-\theta}=R_{J-out}} \left(-\frac{V_{J-\theta}{}^2}{R_{J-\theta}}\cdot\rho h_{J-\theta}\right)R_{J-\theta}dR_{J-\theta} \\ &= -2\pi\int_{R_{J-in}}^{R_{J-out}} V_{J-\theta}{}^2\cdot\rho h_{J-\theta}dR_J \end{aligned} \tag{10a}$$

which is a downward force, or an upward force

$$L_{pressure} = Force_{\substack{Pressure\\CV-lower}} - Force_{\substack{Pressure\\CV-upper}} = 2\pi\int_{R_{J-in}}^{R_{J-out}} V_{J-\theta}{}^2\cdot\rho h_{J-\theta}dR_J \tag{10b}$$

$$L_{pressure} = 2\pi\int_{R_{J-in}}^{R_{J-out}} V_{J-\theta}{}^2\cdot\rho h_{J-\theta}dR_J = \dot{m}\int_{R_{J-in}}^{R_{J-out}} \frac{V_{J-\theta}}{R_J}dR_J = \dot{m}V_{J-in}\left(\ln R_{J-out} - \ln R_{J-in}\right) \tag{10c}$$

Hence, the total lift due to Coandă jet blanket momentum and Coandă jet blanket pressure difference, or the total lift force due to the rate of momentum in the vertical direction and due to pressure difference across the control volume becomes

$$L_{Coanda\,MAV} = L_{\substack{Momentum\\Coanda\,jet\,blanket}} + L_{pressure} = 2\pi R_o h_o\,\rho_J\cdot V_{J.in}^2 + \dot{m}V_{J-in}\ln\frac{R_{J-out}}{R_{J-in}} \tag{11a}$$

or

$$L_{Coanda\,MAV} = \dot{m}V_{J-in}\left(1+\ln\frac{R_{J-out}}{R_{J-in}}\right) \tag{11b}$$

Defining Coandă jet input momentum rate as

$$2\pi R_{in}h_{J-in}\rho_J V_{J.in}{}^2 = \dot{m}_{J.in}V_{J.in} \tag{12}$$

The Coandă jet momentum coefficient (as initially defined by Poisson-Quinton [38]) is conventionally defined as

$$C_\mu = \frac{\dot{m}_{J.in}V_{J.in}}{\frac{1}{2}\rho U^2 S} = \frac{2\pi R_{in}h_{in}\rho_J V_{J.in}{}^2}{\frac{1}{2}\rho U^2 S} \tag{13}$$

http://dx.doi.org/10.5772/intechopen.70157

where U and S should be referred to the ambient reference condition, which here can be chosen to be either the free stream velocity U_∞ and the lifting area of the Coandă device (MAV), πR_o^2, respectively, or the downward inlet velocity to the Coandă MAV jet inducing rotor, i.e., the velocity of the incoming downward flow to the rotor U_R (see **Figure 5**) and the area of the rotor A_R. Hence,

$$C_{\mu-FS} = \frac{\dot{m}_{J.in} V_{J.in}}{\frac{1}{2}\rho U_\infty{}^2 \pi R_{out}{}^2} \tag{14a}$$

or

$$C_{\mu-Rotor} = \frac{\dot{m}_{J.in} V_{J.in}}{\frac{1}{2}\rho U_R{}^2 \pi R_R{}^2} \tag{14b}$$

For rotor in-flow completely dedicated for Coandă jet, from conservation of mass principle

$$\dot{m}_{J.in} = \rho U_R \pi R_R{}^2 \tag{15}$$

and

$$C_{\mu-Rotor} = \frac{\dot{m}_{J.in} V_{J.in}}{\frac{1}{2}\rho U_R{}^2 \pi R_R{}^2} = \frac{2\pi R_{in} h_{in} \rho V_{J.in}{}^2}{\frac{1}{2}\rho U_R{}^2 \pi R_R{}^2} = 4\frac{R_{in} h_{in}}{R_R{}^2}\frac{V_{J.in}{}^2}{U_R{}^2} \tag{16}$$

Hence,

$$C_{\mu-Rotor} = \frac{2\pi R_{J-in} h_{J-in} \rho V_{J.in}{}^2}{\frac{1}{2}\pi R_R{}^2 \rho U_R{}^2} = 4\left(\frac{h_{J-in}}{R_{J-in}}\right)\left(\frac{V_{J.in}}{U_R}\right)^2 \equiv 4\left(\frac{h_{J-in}}{R_R}\right)\left(\frac{V_{J.in}}{U_R}\right)^2 \tag{17}$$

For convenience, in line with the rationale elaborated in [31–34], an indicator of the effectiveness of the Coandă jet device (MAV) to produce lift from rotor-like momentum rate input from the ambient air, a Performance Measure PM can be defined. Following the previous development and different from [31–34], a new Performance Measure P_M can be defined as

$$P_{M-Coandă} = \frac{Lift}{Rate\ of\ Momentum\ Input_{Coandă}} = \frac{\dot{m} V_{J-in}\left(1 + \ln\frac{R_{J-out}}{R_{J-in}}\right)}{2\pi R_{in} h_{in} \rho_J V_{J.in}{}^2} = \left(1 + \ln\frac{R_{J-out}}{R_{J-in}}\right) \tag{18}$$

Utilizing Eq. (16) and assuming $R_{J\text{-}in} \approx R_R$ (without loss of generalities), $C_{\mu\text{-Rotor}}$ can be expressed as

$$C_{\mu-Rotor} = 4\frac{R_{in} h_{in}}{R_R{}^2}\frac{V_{J.in}{}^2}{U_R{}^2} \approx 4\frac{h_{J-in}}{R_{J-in}}\frac{V_{J.in}{}^2}{U_R{}^2} \approx 4\frac{h_{J-in}}{R_R{}^2}\frac{V_{J.in}{}^2}{U_R{}^2} \tag{19}$$

The performance measure (PM) can alternatively be written in terms of momentum coefficient as

$$P_{M-Coandă} = \frac{Lift}{Rate\ of\ Momentum\ Input_{Coandă}} = \frac{\dot{m}V_{J-in}\left(1+\ln\frac{R_{J-out}}{R_{J-in}}\right)}{2\pi R_{J-in}h_{J-in}\rho_J V_{J.in}{}^2}$$
$$= 4\frac{R_{J-in}h_{J-in}}{R_R{}^2}\frac{V_{J.in}{}^2}{U_R{}^2}\frac{\left(1+\ln\frac{R_{J-out}}{R_{J-in}}\right)}{C_{\mu-Rotor}} \tag{20a}$$

or

$$P_{M-Coandă} = \frac{Lift}{Rate\ of\ Momentum\ Input_{Coandă}} = \frac{\dot{m}V_{J-in}\left(1+\ln\frac{R_{J-out}}{R_{J-in}}\right)}{2\pi R_{J-in}h_{J-in}\rho_J V_{J.in}{}^2} \approx 4\frac{h_{J-in}}{R_{J-in}}\frac{V_{J.in}{}^2}{U_R{}^2}\frac{\left(1+\ln\frac{R_{J-out}}{R_{J-in}}\right)}{C_{\mu-Rotor}} \tag{20b}$$

The above derivation is carried out on a Coandă MAV configuration with actuator (rotor) dedicated to produce Coandă Jet Blanket. It has tacitly been assumed that the Coandă MAV jet blanket covers the entire upper surface of the Coandă MAV ("perfect blanket"). Therefore, for this Coandă MAV semi-spherical configuration design, which may represent a baseline of other similar Coandă MAV with moderately different upper surface curvature with MAV with ("perfect blanket"), the wake of the Coandă MAV is shed as a wake following the body of the Coandă MAV. Such situation can be verified at a later stage using CFD simulation for closely modelled flow situation. Then it will be reasonable to assume that the static pressure of the wake of the Coandă MAV the ambient pressure p_a prevails. For Coandă MAV with ducted-fan-line configuration, an actuator disk model should be added in the force system in addition to those elaborated above.

3. Lift generated by the rotor (actuator disk)

Figure7 depicts a MAV configuration consisting of a propeller at the center of the upper part with radially bled flow to generate Coandă effect along the upper surface of the spherical MAV. Here, UR is the axial inlet flow through the actuator along the vertical actuator shaft direction.

Therefore, the lift generated by the propeller can be calculated using actuator disk theory [39–42]

$$Lift_{actuator} = 2\pi\rho R_R{}^2 U_R{}^2 \tag{21}$$

which should be added to Eq. (12), and modified to account for the actuator disk cylindrical part with assumed uniform radius R_J. Therefore, the lift of the Coandă MAV due to the actuator and Coandă blanket becomes

$$Lift_{\substack{Spherical-Coanda\,MAV\\ with-actuator}} = Lift_{\substack{Spherical-Coanda\,MAV\\ without-actuator}} + Lift_{actuator} = \dot{m}V_{J-in}\left(1+\ln\frac{R_{J-out}}{R_{J-in}}\right) + 2\pi\rho R_R{}^2 U_{R\text{-}actuator}{}^2$$
$$= \dot{m}V_{J-in}\left(1+\ln\frac{R_{J-out}}{R_{J-in}}\right) + 2\dot{m}_R U_{R\text{-}actuator} \tag{22}$$

Here, the mass flow rate into the actuator disk is given by

$$\dot{m}_R \equiv \pi\rho R_R{}^2 U_{R\text{-}actuator} \tag{23}$$

and one may assume that

$$U_\infty \approx 0 \tag{24}$$

and from actuator disk theory [40–43]

$$w = 2U_{R\text{-}actuator} \tag{25}$$

In this elaboration, the Betz limit [43], which is the theoretical maximum efficiency of the actuator disk ($\eta = 16/27 \approx 59.26\%$), is not considered.

A remark is in order here associated with the Coandă MAV as illustrated in **Figures. 4**, and **5** as represented and the top of **Figure 7** above. To avoid the MAV body counter-rotates due to the rotor, counter-rotating rotors can be utilized as the actuator. Eqs (13)–(15) remain valid as the baseline equation, but the design configuration in the conceptual and detailed design of the Coandă MAV should be elaborated accordingly.

Several alternatives can be considered for the Coandă MAV with actuator configuration. First, if all the flow into the actuator is utilized for the Coandă MAV, one may obtain a comparison between the lift produced by Coandă MAV without actuator and ducted fan which utilizes actuator flow only. Hence

$$\frac{Lift_{\substack{Spherical-Coanda\,MAV\\ without-actuator}}}{Lift_{actuator}} = \frac{\dot{m}V_{J-in}\left(1+\ln\frac{R_{J-out}}{R_{J-in}}\right)}{2\pi\rho R_R{}^2 U_R{}^2} == \frac{1}{2}\left(\frac{\dot{m}}{\dot{m}_{actuator}}\right)\left(1+\ln\frac{R_{J-out}}{R_{J-in}}\right)\frac{V_{J-in}}{U_{R\text{-}actuator}}$$
$$= \frac{\Phi_{Coandă-mass}}{2}\left(1+\ln\frac{R_{J-out}}{R_{J-in}}\right)\frac{V_{J-in}}{U_{R\text{-}actuator}} \tag{26}$$

Here, the ratio between the mass flow rates injected for Coandă MAV jet $\dot{m}_J$ and the mass flow rate entering the rotor $\dot{m}_R$ is defined as the Coandă MAV mass ratio $\Phi_{\text{Coandă-mass}}$.

$$\Phi_{Coandă-mass} = \frac{\dot{m}_J}{\dot{m}_R} \tag{27}$$

Eq. (26) establishes the benefit of Coandă MAV jet blanket lift compared to actuator (ducted fan, or helicopter) lift,

Eq. (22) can be written as

$$Lift_{\substack{\text{Spherical - Coanda MAV}\\ \text{with - actuator}}} = \dot{m}_J V_{J-in}\left(1+\ln\frac{R_{J-out}}{R_{J-in}}\right) + 2\pi\rho R_R{}^2 U_{R\text{-actuator}}{}^2 = \dot{m}_J V_{J-in}\left(1+\ln\frac{R_{J-out}}{R_{J-in}}\right) + 2\dot{m}_R U_{R\text{-actuator}}$$
$$= \dot{m}_R U_{R\text{-actuator}}\left\{\Phi_{Coand\breve{a}-mass}\frac{V_{J-in}}{U_{R\text{-actuator}}}\left(1+\ln\frac{R_{J-out}}{R_{J-in}}\right)+2\right\} \tag{28}$$

Then the performance measure (PM) for Lift can alternatively be written in terms of Coandă MAV mass ratio $\Phi_{Coanda-mass}$ as

$$P_{M-Lift} \equiv \frac{Lift_{\substack{\text{Spherical - Coanda MAV}\\ \text{with - actuator}}}}{\dot{m}_R U_{R\text{-actuator}}} = \left\{\Phi_{Coanda-mass}\frac{V_{J-in}}{U_{R\text{-actuator}}}\left(1+\ln\frac{R_{J-out}}{R_{J-in}}\right)+2\right\} \tag{29}$$

If all the actuator flow is utilized for the Coandă MAV jet blanket, Eq. (28) reduces to

$$Lift_{\substack{\text{Spherical - Coanda MAV}\\ \text{with - actuator}}} = \dot{m}_R\left\{V_{J-in}\left(1+\ln\frac{R_{J-out}}{R_{J-in}}\right)\right\} \tag{30}$$

and the performance measure (PM) for Lift of Coandă MAV Eq. (29) reduces to

$$P_{M-Lift} \equiv \frac{Lift_{\substack{\text{Spherical - Coanda MAV}\\ \text{with - actuator}}}}{\dot{m}_R U_{R\text{-actuator}}} = \left\{\frac{V_{J-in}}{U_{R\text{-actuator}}}\left(1+\ln\frac{R_{J-out}}{R_{J-in}}\right)+2\right\} \approx \left\{\frac{R_{J-in}}{2h_{J-in}}\left(1+\ln\frac{R_{J-out}}{R_{J-in}}\right)+2\right\} \tag{31}$$

4. Flight dynamic analysis for semi-spherical Coandă MAV in translatory motion

4.1. System of coordinates

For setting the equation of motion of the Coandă MAV, the system coordinates that are required will be defined. Four main reference frames can be identified for this purpose; these are the inertial coordinate system, the local horizon reference frame, the body reference frame, and the wind reference frame. These reference frames are shown in **Figure 8**.

For simplicity and instructiveness, only two-dimensional coordinate systems in the plane of symmetry of the Coandă MAV will be defined. These four frame of reference are:

1. **Inertial coordinate system**, which is used for defining and the application of the Newton's law of motion. In two dimensions, this coordinate system is depicted by the $O_E xyz$.

2. **The local horizon coordinate system $Ox_hy_hz_h$**, which is fixed at the center of mass of the vehicle, and is parallel with the earth Inertial Coordinate system.

3. **The body coordinate system $Ox_by_bz_b$**, which is fixed to the vehicle and follows a conveniently chosen axis of the vehicle.

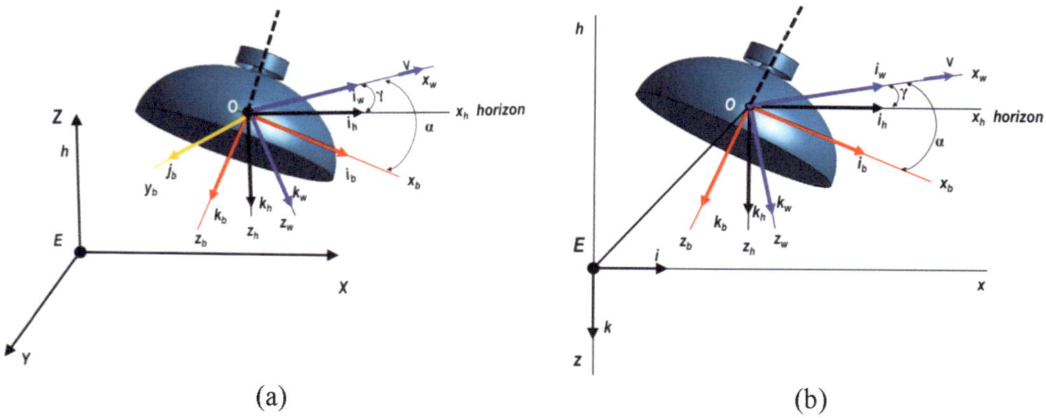

Figure 8. (a) The details of the four coordinate systems for Coandă MAV flight vehicle; (b) Axes systems in the vertical plane perpendicular to the Earth's Surface.

4. **The wind axes system $Ox_wy_wz_w$,** which moves with the vehicle and the x_w axis coincides with the flight path of the vehicle, hence its velocity vector.

The wind axes are oriented to the flight path angle γ relative the local horizon axes, as shown in **Figure 8a**, and by the angle of attack α relative to the body axes. **Figure 8b** exhibits the two dimensional configuration in the vertical plane perpendicular to the earth surface.

4.2. Flight dynamic analysis

Without loss of generalities, the equation of motion of the flight dynamics of Coandă MAV in translatory motion is developed in the vertical plane perpendicular to the earth motion (two-dimensional planar motion). The hovering state, illustrated in **Figure 9**, is employed as a reference. In addition, in the present analysis, only the Coandă MAV jet is incorporated in the

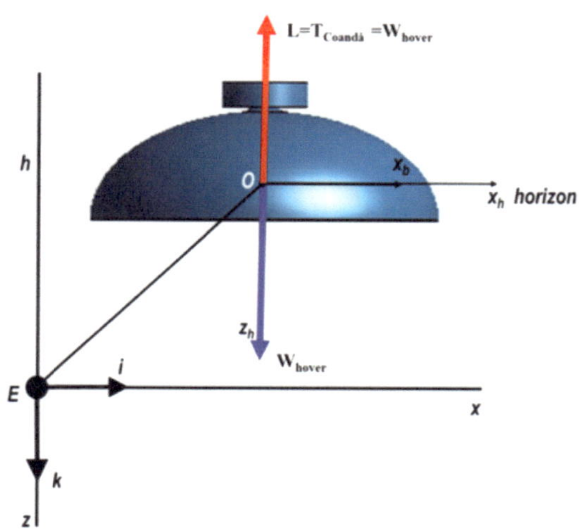

Figure 9. Coordinate systems for Coandă MAV in hover, in two-dimensional flight (at the vertical plane of symmetry of the Coandă MAV).

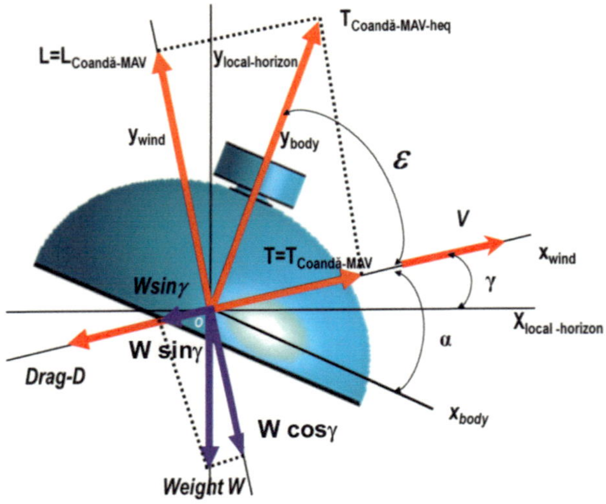

Figure 10. Coandă MAV schematic in two-dimensional translatory motion in the vertical plane (at the vertical plane of symmetry and motion).

derivation. The influence of the actuator can be linearly superposed as demonstrated in the hover case, Eqs. (13)–(15).

Balance of forces in the free-body diagram exhibited in **Figure 10** implies

$$L_{Hover} \equiv T_{Coandă-MAV} = W_{Coandă-MAV-Take-off} \tag{32}$$

The use of the mass conservation equation

$$\dot{m}_{J-in} = 2\pi\rho R_{J-in} h_{J-in} V_{J\text{-}in} \tag{33}$$

and the use of the Coandă lift from Eq. (11b) then leads to:

$$L_{Hover} \equiv T_{Coandă-MAV} = 2\pi\rho R_{J-in} h_{J-in} V_{J\text{-}in}^2 \left(1 + \ln\frac{R_{J-out}}{R_{J-in}}\right) = W_{Coandă-MAV-Take-off} \tag{34}$$

Further, simplifying assumptions will be made for the development of the equation of motion during translatory motion. This equation could be further refined to incorporate more realistic cases.

These comprise

1. During hover, the thrust of the Coandă MAV will not be influenced by the inclination of the Coandă MAV. This assumption can be later modified for more elaborate analytical model, which could be assisted by CFD simulation.

2. The thrust generated by the Coandă MAV during translatory motion is assumed to stay the same; therefore:

$$\begin{aligned} T_{Coand\breve{a}-MAV} &= \dot{m}_{J-in} V_{J-in}\left(1+\ln\frac{R_{J-out}}{R_{J-in}}\right)|_{hover-equivalent} \\ &= 2\pi\rho R_{J-in} h_{J-in} V_{J-in}^2\left(1+\ln\frac{R_{J-out}}{R_{J-in}}\right)|_{hover-equivalent} \end{aligned} \quad (35)$$

Using the basic flight dynamic equation for flight vehicle, the equation of motion of the Coandă MAV in translational motion in the two-dimensional vertical plane can be written by referring to **Figure 10**

Kinematic:

$$\dot{x} = V\cos\gamma \quad [LT^{-1}] \quad (36)$$

$$\dot{h} \equiv \dot{y} = V\sin\gamma \quad [LT^{-1}] \quad (37)$$

$$\varepsilon = \tan^{-1}\frac{\ddot{y}_{wind}}{\dot{V}} = \tan^{-1}\frac{g(L-W\cos\gamma)}{W\dot{V}} = \tan^{-1}\frac{g}{\dot{V}}\frac{(L-W\cos\gamma)}{W} \quad [-] \quad (38)$$

$$\dot{\varepsilon} = \tan^{-1}\frac{\ddot{y}_{wind}}{V} = \tan^{-1}\frac{g(L-W\cos\gamma)}{WV} = \tan^{-1}\frac{g}{V}\frac{(L-W\cos\gamma)}{W} \quad [T^{-1}] \quad (39)$$

Coandă MAV Fluid Dynamics:

$$T_{Coand\breve{a}-MAV} = 2\pi\rho R_{J-in} h_{J-in} V_{J-in}^2\left(1+\ln\frac{R_{J-out}}{R_{J-in}}\right) \quad [MLT^{-2}] \quad (40)$$

Coandă MAV Flight Dynamics:

$$\dot{V} = (g/W)\left(T_{Coand\breve{a}\ MAV-heq}\cos\varepsilon - D - W\sin\gamma\right) \quad [LT^{-2}] \quad (41)$$

$$\ddot{y} = (g/W)\left(T_{Coand\breve{a}\ MAV-heq}\sin(\varepsilon+\gamma) - W - D\sin\gamma\right) \quad [LT^{-2}] \quad (42)$$

Weight-Fuel Consumption:

$$\dot{W} = -CT \quad [MT^{-1}] \quad (43)$$

where C is the specific fuel consumption, weight per unit thrust [Newton/Newton].

The above relationships are derived for balance of forces and Newton's equation for semi-spherical Coandă MAV, which is treated as a point mass moving in a plane perpendicular to the planar Earth.

5. Performance measure

5.1. Performance measure during hover

The feasibility of using Coandă techniques to enhance aerodynamic performance of Coandă MAV can be outlined through some non-dimensional quantities of performance measure. The most logical measure that has commonly been utilized is the Coandă jet momentum coefficient, C_{μ}(Poisson-Quinton [38]), which are defined as (14a) or (14b), as appropriate:

$$C_{\mu-\text{Coandă}-jet} = \frac{\dot{m} V_{\text{Coandă}-jet}}{\frac{1}{2}\rho_j V_\infty^2 A_{ref}} \tag{44a}$$

or

$$C_{\mu-\text{Coandă}-jet} = \left[\frac{\dot{m}_{J.in} V_{J.in}}{\frac{1}{2}\rho U_\infty{}^2 \pi R_{out}{}^2}\right]_{\text{Coandă}-jet} \tag{44b}$$

or

$$C_{\mu-\text{Coandă}-jet} = \left[\frac{\dot{m}_{J.in} V_{J.in}}{\frac{1}{2}\rho U_\infty{}^2 \pi R_{out}{}^2}\right]_{\text{Coandă}-jet} \tag{44c}$$

One can define another Performance Measure to evaluate the aerodynamic performance of spherical Coandă MAV considered here during hover and lift-off, by assessing the output lift as compared to the input momentum rate, following consideration discussed in [31, 33–34]. These are given by Eqs. (18) and (20) following a different derivation elaborated in previous section.

The Performance Measure may be defined in a different way by referring to the rate of the energy of the Coandă jet ejected at its ideal inlet velocity $V_{j\text{-}out}$ at its peripheral outlet, and comparing it to the rate of momentum influx of the Coandă jet times its inlet velocity $V_{j\text{-}in}$ (non-dimensional):

$$P_{\substack{M-E \\ \text{Spherical Coandă}-MAV}} = \frac{\textit{Lift produced by Coandă MAV} \text{ times } V_{J-out} \textit{ at its peripheral outlet}}{\textit{the rate of momentum influx of the Coandă jet times its inlet velocity } V_{J-in}}$$

$$= \frac{\dot{m} V_{J-in} V_{J-out}\left(1 + \ln\frac{R_{J-out}}{R_{J-in}}\right)}{2\pi R_{in} h_{in} \rho_J V_{J.in}{}^3} = \left(1 + \ln\frac{R_{J-out}}{R_{J-in}}\right)\frac{V_{J-out}}{V_{J.in}} \tag{45}$$

based on Coandă MAV geometry and kinematics, or

$$P_{\substack{M-E \\ \text{Spherical Coandă}-MAV}} = \frac{\textit{Lift produced by Coandă MAV} \text{ times } V_{J-out} \textit{ at its peripheral outlet}}{\textit{the rate of momentum influx of the Coandă jet times its inlet velocity } V_{J-in}}$$

$$\approx 4\frac{h_{J-in}}{R_{J-in}}\frac{V_{J.in}{}^2}{U_R{}^2}\frac{\left(1+\ln\frac{R_{J-out}}{R_{J-in}}\right)}{C_{\mu-Rotor}}\frac{V_{J-out}}{V_{J.in}} \tag{46}$$

based on Coandă MAV geometry, kinematics and momentum coefficient. Although Eq. (18) is based on rate of momentum flux while Eq. (45) on energy input, both equations are similar.

For the baseline semi-spherical Coandă MAV case considered, $V_{J\text{-}out}/V_{J\text{-}in}$ will be the same, and hence Eqs. (45) and (46) reduces to Eqs. (18) and (20). This notion may not be met in more general cases. Since the theoretical analysis does not consider viscosity and other losses, more information is required to insure uniqueness. Further realistic assumption can be drawn by carrying out computational fluid dynamic (CFD) simulation and visualization, which incorporate viscous effects and turbulence, to add further understanding into the problem and introduce further refinement in the theoretical modeling and analysis.

5.2. Performance measure during translation

For further reference and as a baseline, Performance Measure can be defined for two cases during translation: for level flight and for climbing.

For level flight (by referring to **Figure 11**):

$$\begin{aligned} P_{M\,\underset{LevelFlight}{\text{Coandă}-MAV}} &= \frac{\textit{Power required to overcome Drag}}{\textit{Power Input by Coandă jet action}} \\ &= \frac{\textit{Drag} * \textit{Velocity}}{\textit{Coandă Jet}\ \text{Momentum Gain} * \textit{Coandă Jet}\ \text{Velocity In}} \\ &= \frac{D\cdot V\cos\gamma^2}{\dot{m}V_{j\text{-}in}{}^2}\left[\frac{ML^2T^{-2}}{ML^2T^{-2}}\right] \end{aligned} \tag{47}$$

For purely level flight, $\gamma = 0$ (by referring to **Figures 10** and **11**). Taking into account that the prevailing cross-sectional area of the Coandă MAV perpendicular to the flow is depicted in **Figure 11(b)**, Eq. (47) reduces to:

$$P_{M\,\underset{LevelFlight}{\text{Coandă}-MAV}} \approx \frac{1}{4}C_{D\to x}\frac{\pi\rho R_o{}^2 sin\alpha}{\dot{m}_i}\frac{V^3}{V_{j\text{-}in}{}^2} \tag{48}$$

Alternatively, another related Performance Measure can be defined as follows:

$$P'_{M\,\underset{LevelFlight}{\text{Coandă}-MAV}} \equiv \frac{P_{M\,\underset{LevelFlight}{\text{Coandă}-MAV}}}{C_{D\to x} sin\alpha} \approx \frac{1}{4}\frac{\pi\rho R_o{}^2}{\dot{m}_i}\frac{V^3}{V_{j\text{-}in}{}^2} \tag{49}$$

Here $C_{D\to x}$ is the Drag coefficient of the Coandă MAV in the x direction, and α is its inclination angle with respect to the flight direction as depicted as depicted in **Figure 11**.

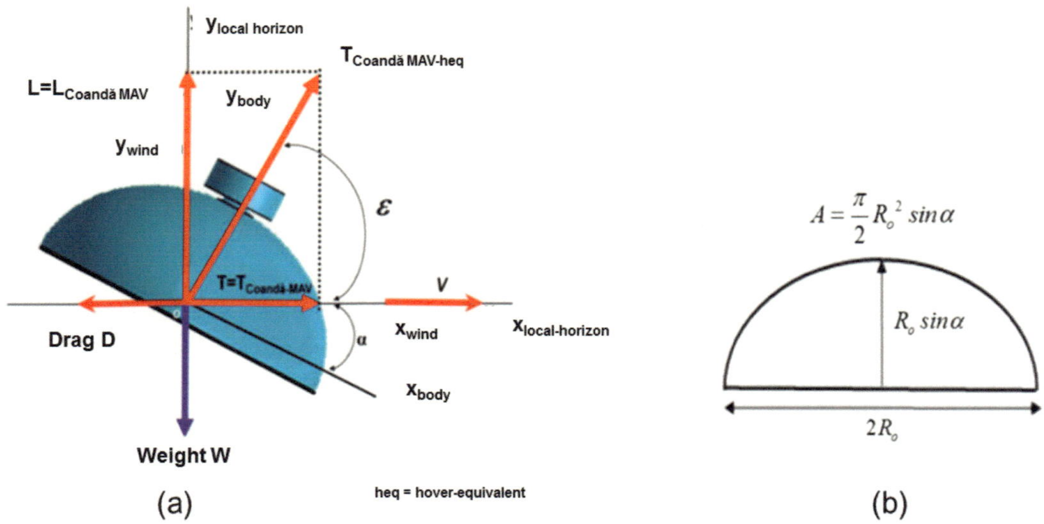

Figure 11. Schematic of Coandă MAV in two-dimensional cruising motion in the vertical plane of symmetry and motion. Inset: Cross-sectional area perpendicular to direction of motion.

The general expression for the Performance Measure for climbing is:

$$\begin{aligned} P_{M\,\substack{\text{Coandă}\\ -MAV \\ \text{Climbing}}} &= \frac{\textit{Power required to Climb}}{\textit{Power Input by Coandă jet action}} \\ &= \frac{T_{\text{Coandă}-MAV} \bullet \dot{y} - D \bullet \dot{y} - W \bullet \dot{y}}{\text{Coandă } \textit{Jet Rate of } \text{Momentum Gain} * \text{Coandă } \textit{Jet } \text{Velocity In}} \\ &= \left(\frac{T(t) - (D \sin\gamma + W)}{\dot{m} V_{J\text{-}in}} \right) \frac{V \sin\gamma}{V_{J\text{-}in}} \end{aligned} \tag{50}$$

Here, all variables are taken for the same flow and flight condition.

For purely climbing flight, as schematically depicted in **Figure 12**, $\gamma = \pi/2$, and Eq. (45) reduces to

$$\begin{aligned} P_{M\,\substack{\text{Coandă}-MAV \\ \text{Climbing}}}\Bigg|_{\substack{\text{based}-\text{on} \\ \text{Coandă Jet} \\ \text{at}-\text{hover}}} &= \frac{T_{\text{Coandă}-MAV} \bullet \dot{y} - D \bullet \dot{y} - W \bullet \dot{y}}{\left(\text{Coandă } \textit{Jet } \ \textit{Rate of } \text{Momentum Gain} * \text{Coandă } \textit{Jet } \text{Velocity In}\right)\Bigg|_{\substack{\text{Coandă Jet} \\ \text{at}-\text{hover}}}} \\ &= \left(\frac{T - (D + W)}{\dot{m} V_{J\text{-}in}} \right) \frac{V}{V_{J\text{-}in}} \end{aligned} \tag{51}$$

Considering the fuel consumed, Eq. (46) becomes

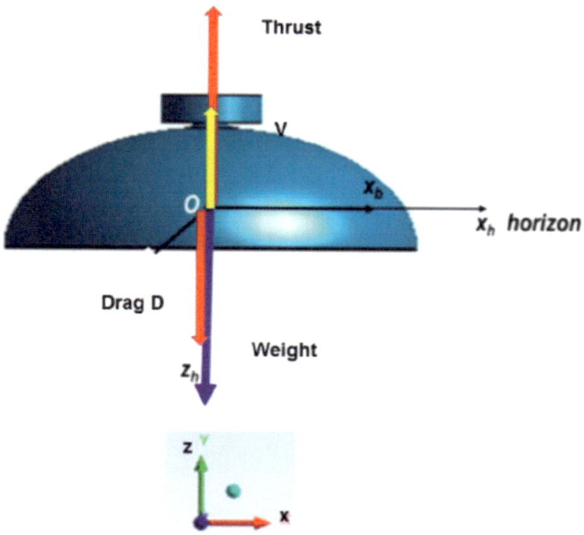

Figure 12. Schematic of Coandă MAV in two-dimensional climb in the vertical plane of symmetry and motion.

$$
\begin{aligned}
P_{M\,\substack{\text{Coandă}-MAV\\ Climbing}}(t)\Big|_{\substack{based-on\\ \text{Coandă } Jet\\ at-hover}} &= \frac{T(t)\bullet\dot{y} - D\bullet\dot{y} - \left(W - \int CT(t)dt\right)\bullet\dot{y}}{(\text{Coandă } Jet\ \ Rate\ of\ \text{Momentum Gain} * \text{Coandă } Jet\ \text{Velocity In})_{hover}} \\
&= \left\{\frac{T(t) + \int CT(t)dt}{\dot{m}V_{\substack{J\text{-}in\\ hover}}} - \left(\frac{C_D}{4}\frac{{R_{J-out}}^2}{R_{J-in}h_{J-in}}\left(\frac{V}{V_{\substack{J\text{-}in\\ hover}}}\right)^2 + \left(1 + \ln\frac{R_{J-out}}{R_{J-in}}\right)\right)\right\}\frac{V}{V_{\substack{J\text{-}in\\ hover}}}.
\end{aligned}
\tag{52}
$$

If we define that, the thrust T of the Coandă MAV should be able to produce climbing motion for a period of θ seconds, straight forward arithmetic calculation using Eq. (47) will require that the propulsion of the Coandă MAV should have a power capacity exceeding

$$
P_{M\,\substack{\text{Coandă}-MAV\\ Climbing}}(t)\Big|_{\substack{based-on\\ \text{Coandă } Jet\\ at-hover}} = \left\{\left(\Gamma + \frac{\Theta}{7200}\right)\left(1 + \ln\frac{R_{J-out}}{R_{J-in}}\right) - \left(\frac{C_D}{4}\frac{{R_{J-out}}^2}{R_{J-in}h_{J-in}}\left(\frac{V}{V_{\substack{J\text{-}in\\ hover}}}\right)^2\right)\right\}\frac{V}{V_{\substack{J\text{-}in\\ hover}}} \geq 0
\tag{53}
$$

corresponding to providing thrust of

$$
T_{\text{Coandă}-MAV} = \left(2\pi\rho R_{J-in}h_{J-in}V_{J\text{-}in}^2\left(1 + \ln\frac{R_{J-out}}{R_{J-in}}\right)\right)_{hover}(1 + \Gamma)
\tag{54}
$$

where

$$\Gamma \geq \frac{\left(\frac{C_D}{4}\frac{R_{J-out}{}^2}{R_{J-in}h_{J-in}}\left(\frac{V}{V_{\substack{J\text{-}in\\hover}}}\right)\right)^2}{\left(1+\ln\frac{R_{J-out}}{R_{J-in}}\right)} - \frac{CT}{W}\Theta \tag{55}$$

These expressions comprise further development of those given in [31, 33–34]. The Performance Measure $P_{MCoand\breve{a}\text{-}MAV\ Level\ Flight}$ can be expressed as a function of C_D,R_o, $\dot{m}$ h_i,α, V, $V_{J\text{-}in}$. These basic relationships can further be altered to address other flight conditions.

6. Results and examples

6.1. Results and examples for hover

The theoretical prediction for the performance measures P_M as given by Eqs. (41) and (42) are depicted in **Figures. 13–16**. Comparison of the analytical model and the CFD simulation results for another Performance Measure P_{M2} defined as the Coandă MAV lift per Coandă jet inlet velocity as suggested in [11, 29] is shown in **Figure 17** and elaborated in [31, 33–34]. The Performance Measure P_{M2} is calculated for various Jet slot thickness, and serve as an assessment and validation of the theoretical model. Certainly the CFD model for the same Coandă MAV configuration is based on Reynolds averaged Navier-Stokes equations which take into account viscous effects, which is not considered in the theoretical model and analysis. Thus, some discrepancies will appear, and the comparison should be judged as such.

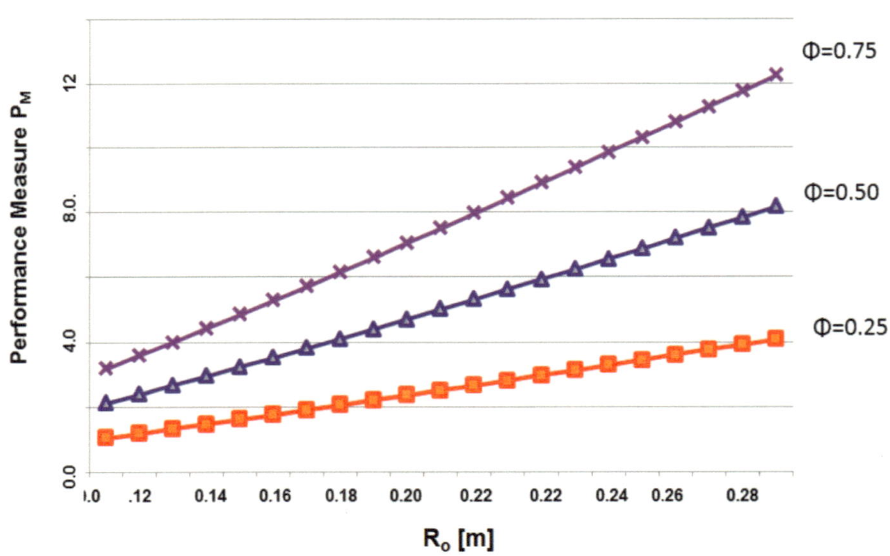

Figure 13. Performance Measure for spherical Coandă MAV for various values of Outer Radius while keeping the Inner Radius R_i and inlet jet slot thickness h_i constant during hover (baseline). In the present example, the Coandă Jet inlet Radius $R_{J\text{-}in} = 0.02$ m, Coandă jet slot thickness $h_{J\text{-}in} = 0.01$ m, and Coandă jet inlet velocity $V_{J\text{-}in} = 5$ m/sec. The performance curves are evaluated at some selected values of mass-flow-Rate Ratio $\Phi_{Coand\breve{a}-mass} = \frac{\dot{m}_J}{\dot{m}_R}$.

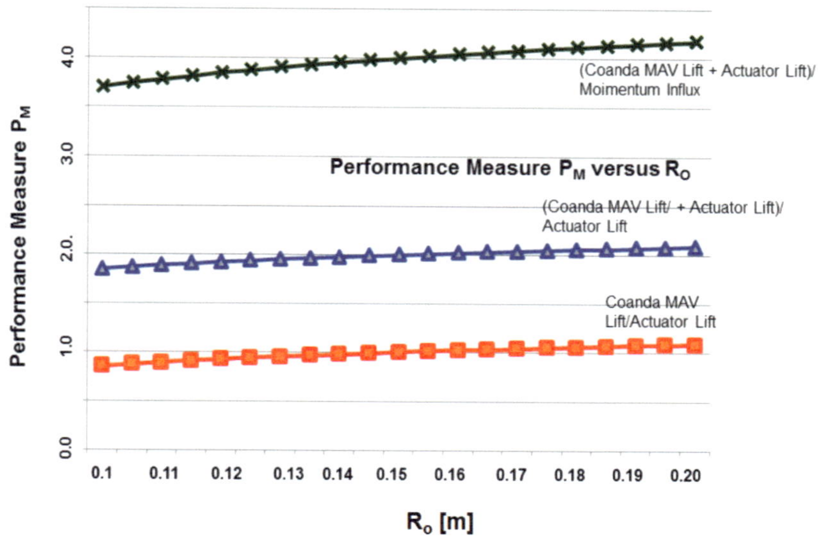

Figure 14. Performance Measure for spherical Coandă MAV for various Outer Radius $R_o[m]$. In the present example, Three different Performance versions are exhibited: Coandă MAV to Actuator Lift Ratio, Combined Coandă MAV and Actuator to Actuator lift Ratio, and Combined Coandă MAV and Actuator to Momentum Rate Ratio. Coandă Jet inlet Radius $R_{J\text{-}in} = 0.02$ m, Coandă jet slot thickness $h_{J\text{-}in} = 0.01$ m, and Coandă jet inlet velocity $V_{J\text{-}in} = 5$m/sec.

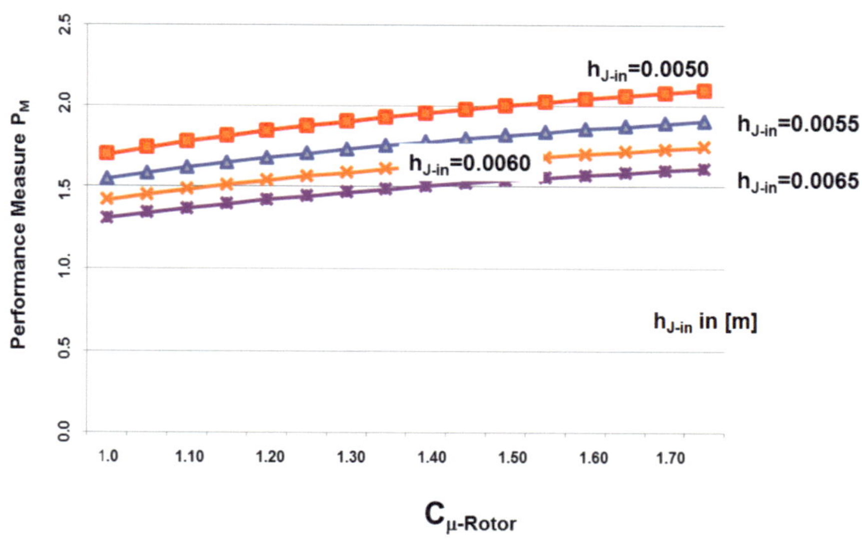

Figure 15. Performance Measure for spherical Coandă MAV as a function of Coandă jet Momentum Coefficient C_μ for various values of $h_{J\text{-}in}$. Similar to **Figure 15**, here the Coandă Jet inlet Radius $R_{J\text{-}in} = 0.02$ m, Coandă jet slot thickness $h_{J\text{-}in} = 0.01$ m, and Coandă jet inlet velocity $V_{J\text{-}in} = 5$ m/sec have been used as the baseline.

Comparison of analytical and CFD [32] models for the Performance Measure P_{M2} defined by total lift L_{MAV}/ Inlet Velocity $V_{j\text{-}in}$ for spherical Coandă is depicted in **Figure 17**. The figure clearly exhibits the noticeable influence of the ratio of the jet slot thickness h to the reference radius R_o, to the Performance Measure P_{M2}. The influence of the jet inlet velocity on lift enables designers to appropriately select the right propulsion system specifications for the Coandă MAV.

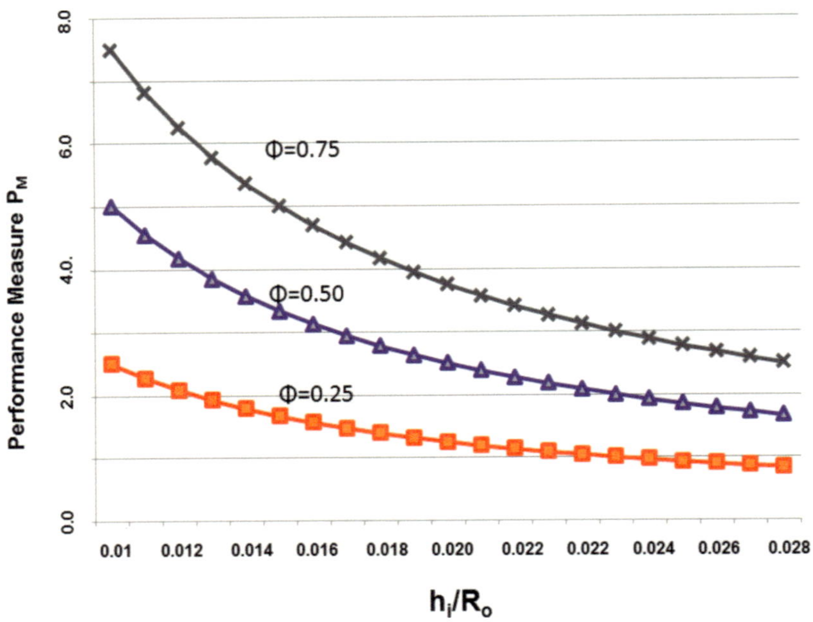

Figure 16. Performance Measure for spherical Coandă MAV as a function of $h_{J\text{-}in}$ /R_o for various values of Coandă jet mass-flow-Rate Ratio $\Phi_{Coandă-mass} = \frac{\dot{m}_J}{\dot{m}_R}$. Similar to **Figure 15**, here the Coandă Jet inlet Radius $R_{J\text{-}i\ n}$ = 0.02m, Coandă jet slot thickness $h_{J\text{-}in}$=0.01m, and Coandă jet inlet velocity $V_{J\text{-}in}$=5m/sec.

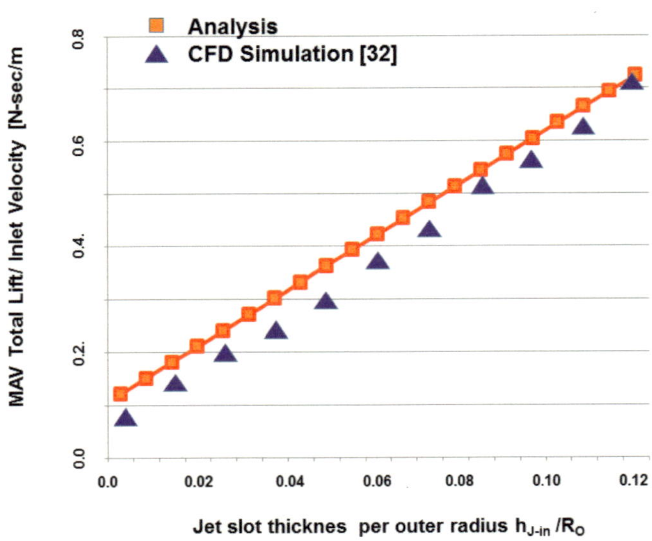

Figure 17. Analytical and CFD [32] models comparison Total Lift L_{MAV}/Inlet Velocity $V_{j\text{-}in}$ for spherical Coandă MAV for various Jet slot thickness $h_{j\text{-}in}$ per outer radius R_o.

Figure 18 exemplifies the effect of viscosity for two different locations of the Coandă jet introduction that can be revealed by CFD simulation. The CFD images in **Figure 18 (a)** and **(b)**, which simulate the velocity magnitude contours for similar configurations considered by Mirkov and Rasuo [11, 29] and Ghozali [30], show the velocity magnitude contours for each of these configurations and exhibit the influence of the inlet configuration. The resulting lift as a function of input jet velocity diagram for each of these configurations are also shown. The corresponding computational grid for each of the configurations utilized by Mirkov and Rasuo, and Ghozali, is also shown as inset in each figure.

For these CFD simulations, which were performed using steady RANS (Reynolds Averaged Navier-Stokes), two equations k-ω shear stress transport (SST) turbulence model is employed. In the solution procedure followed, simple pressure-velocity coupling scheme with the least squares cell base as discretization gradient has been applied; second order upwind for the momentum equation and the turbulent kinetic energy were also imposed.

Figure 18 also exhibits structured numerical mesh used in the flow domain. Through meticulous attempts and grid sensitivity studies the size of the mesh cells was generated small to enable visualizations of the flow around the whole body with best details. The CFD simulations are carried out on two configuration versions: in **Figure 18 (a)**, the flow configuration and grid mesh are closely similar to those utilized by Mirkov and Rasuo [11, 29], and in **Figure 18 (b)**, the flow configuration and grid mesh are closely similar to those utilized by Ghozali [30]. The inset of each figure shows the mesh details adopted by Mirkov and Rasuo [29], and Ghozali [30], respectively. Moderately small Reynolds number, $Re = 68458$, based on mean velocity and jet inlet height has been employed in the numerical study. Additionally, across the inlet which has a thickness of 0.05 m, the velocity is st to be uniform ($V_{j\text{-}in} = 20$ m/s). The CFD simulations utilize computational grid with 52,830 mesh cells. Assessment of the results and referring to the procedure guidelines assure that the mesh quality was acceptable. The CFD computational results indicate that orthogonality quality was 4.96943e-01. Again, from the ANSYS orthogonality quality requirements, these figures should be acceptable. Further details are elaborated by Djojodihardjo et al. [31, 32–35] and Ahmed et al. [33, 34–36].

Grid sensitivity [18–19] and iteration convergence CFD simulation [44] studies were carried out and exhibited in **Figure 19**. The mesh quality and comparable agreements between the present CFD simulation results exhibited in **Figure 18**, as well as the supporting grid mesh and convergence studies exemplified by **Figure 19** demonstrates that the generation of mesh in the present CFD simulations meet the present quality requirements as well as those stipulated in the ANSYS™© guidelines; therefore these can be considered to be acceptable.

The CFD simulations are also carried out to investigate the influence of the jet inlet radius on the air vehicle performance (lift). The CFD simulations were carried out for two inlet jet radius, $R_i = 5$ mm and $R_i = 50$ mm for the baseline semi-spherical Coandă MAV configurations studied. The results are presented as velocity magnitude contours, which exhibit the influence of the inlet radius on the flow field. **Figures 20** and **21** exhibits the pressure contours along semi-spherical Coandă MAV surface obtained by CFD simulation results and the static pressure associated with it in color-coded contour. **Figure 22** compares the CFD simulation results for the pressure coefficient of on the Coandă MAV surface jet blanket obtained from the

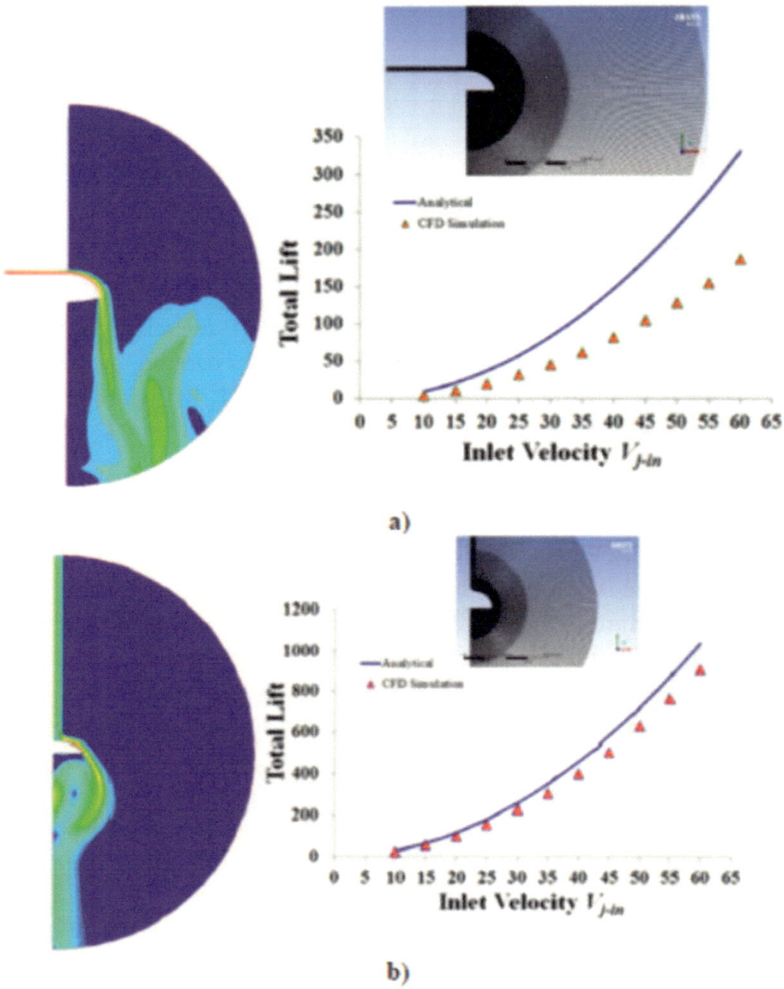

Figure 18. Lift versus input jet velocity for similar configuration considered in (a) Mirkov and Rasuo [11, 29], (b) Ghozali [30]. The inset of each figure show the mesh details adopted by Mirkov and Rasuo [29], and Ghozali [30], respectively, that have been reproduced in the present CFD simulations.

present CFD simulation and those obtained by Mirkov and Rasuo [29]. Overall, these results serve to verify the plausibility of the analytical as well as the CFD simulation efforts.

6.2. Results and examples for translation

For translatory flight, some examples of Performance Measures simulation studies for Coandă MAV in purely level are exhibited by **Figures 23** and **24**, obtained using Eqs. (44) and (48), respectively.

Figure 23 is worked out for the Coandă Jet inlet Radius $R_{J\text{-}in} = 0.02$ m, Coandă jet slot thickness $h_{J\text{-}In} = 0.01$ m, and Coandă jet inlet velocity $V_{J\text{-}n} = 5$ m/sec.. **Figure 24** shows the Thrust ratio Γ as a function of versus flight velocity for pure climbing, following Eq. (49) with equal sign,

http://dx.doi.org/10.5772/intechopen.70157

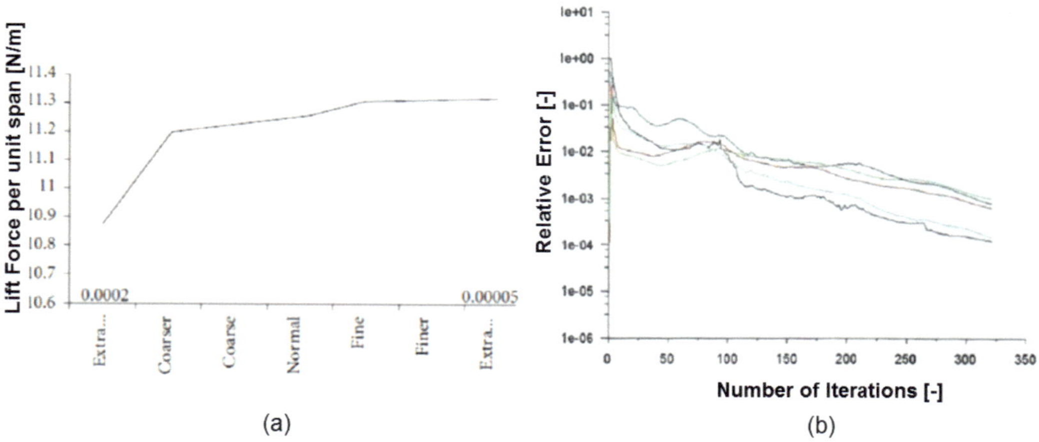

Figure 19. (a) Two-dimensional CFD grid sensitivity study for lift force per unit span on an airfoil at 0° angle of attack [18–19]; (b) Iteration Convergence CFD simulation studies [44]. Both are carried out using ANSYS® [45]. The curves are untouched computer outputs.

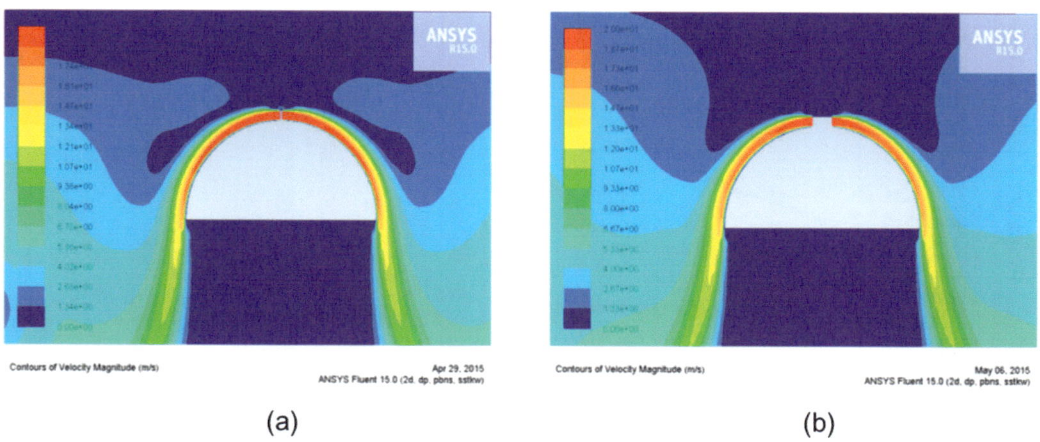

Figure 20. CFD simulation on two Coandă MAV configurations (a) Ri=5 mm, (b) $R_i = 50$ mm. Relatively symmetrical Coandă effect velocity magnitude contours are exhibited. The images are untouched computer outputs.

assuming $C_D \approx 2.0$ and specific fuel consumption C = 0.000139 and period = 1 h. The example is also worked out for the Coandă Jet inlet Radius $R_{J\text{-}in} = 0.02$ m, Coandă jet slot thickness $h_{J\text{-}in} = 0.01$ m, and Coandă jet inlet velocity $V_{J\text{-}in} = 5$ m/sec.

Parametric study results exhibited by these figures can be utilized to give insight on the performance of the semi-spherical Coandă MAV considered as well as for preliminary sizing purposes.

7. Concluding remarks

A comprehensive and meticulous analysis has been made to analyze the potential utilization of Coandă MAV as an aerial robotic platform (vehicle) and deriving the Coandă MAV

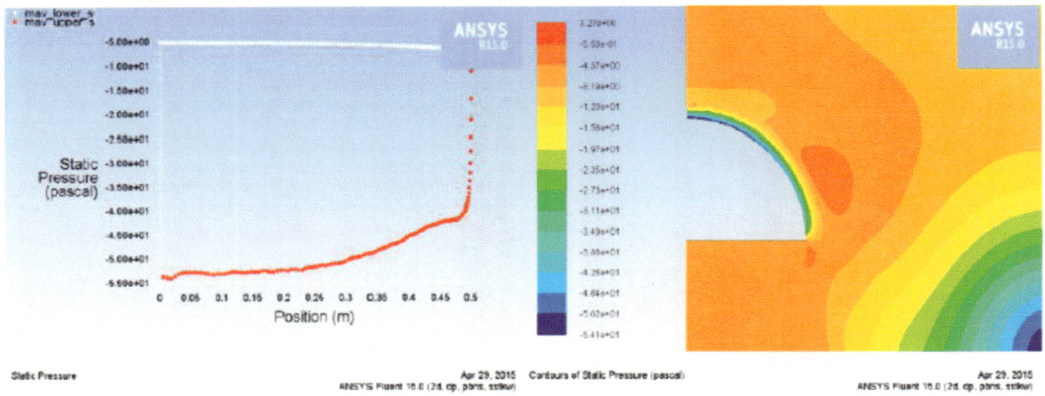

(a) (b)

Figure 21. CFD simulation results plot of surface static pressure (a) and flow field pressure contours image (b) for the baseline semi-spherical Coandă MAV. The images are untouched computer outputs.

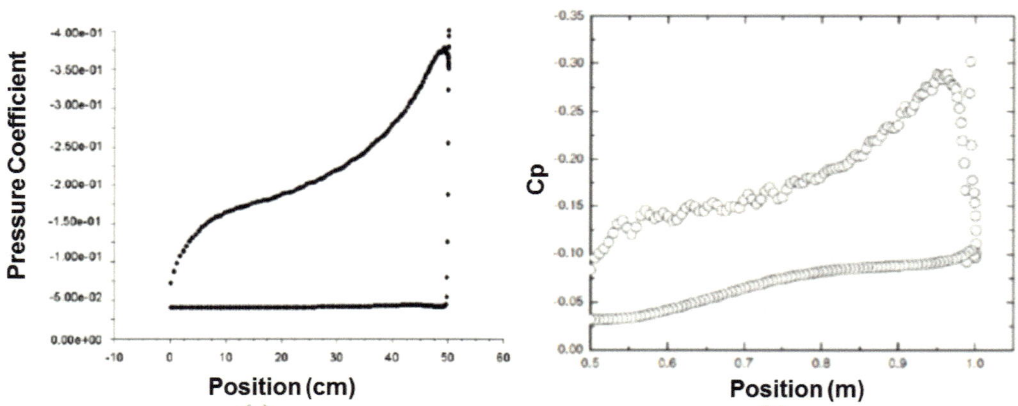

(a) (b)

Figure 22. CFD computational results comparison of the surface pressure coefficient on the Coandă MAV jet blanket; (a) presents CFD simulation results and (b) Mirkov and Rasuo CFD simulation results [11, 29]. The images are untouched computer outputs.

governing equations. Utilizing first principles, i.e., conservation principles for control volume analysis as baseline fluid-dynamic analysis for the Coandă MAV in hover and free-body diagram and Newton's law of motion as baseline flight-dynamic analysis for in-plane translatory motion, the analysis establishes the basic working relationships among various relevant variables and parameters governing the aerodynamics forces and performance measures of Coandă MAV in hover and translatory motion.

To gain insight in the development of the analytical model, and to assess and validate the analytical model subject to the theoretical assumptions, CFD simulations for the Coandă MAV generic models are carried out. Comparison of the theoretical analysis and CFD simulation

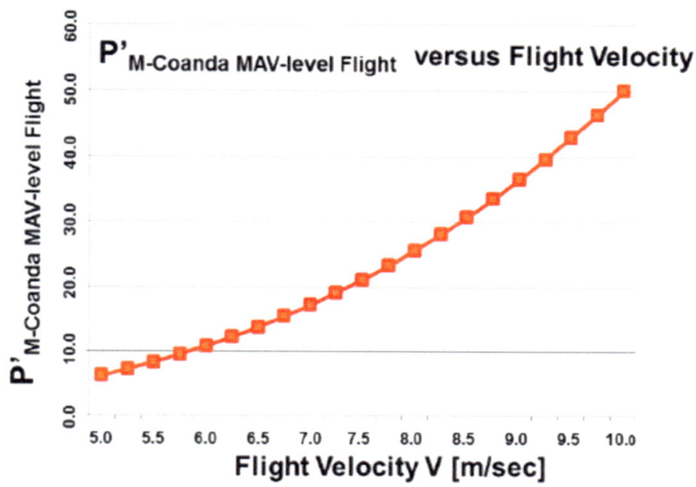

Figure 23. Coandă MAV Performance Measure for level flight as a function of flight velocity as defined by Eq. (44a). The example is worked out for the Coandă Jet inlet Radius $R_{J\text{-}in} = 0.02$ m, Coandă jet slot thickness $h_{J\text{-}in} = 0.01$ m, and Coandă jet inlet velocity $V_{J\text{-}in} = 5$ m/sec.

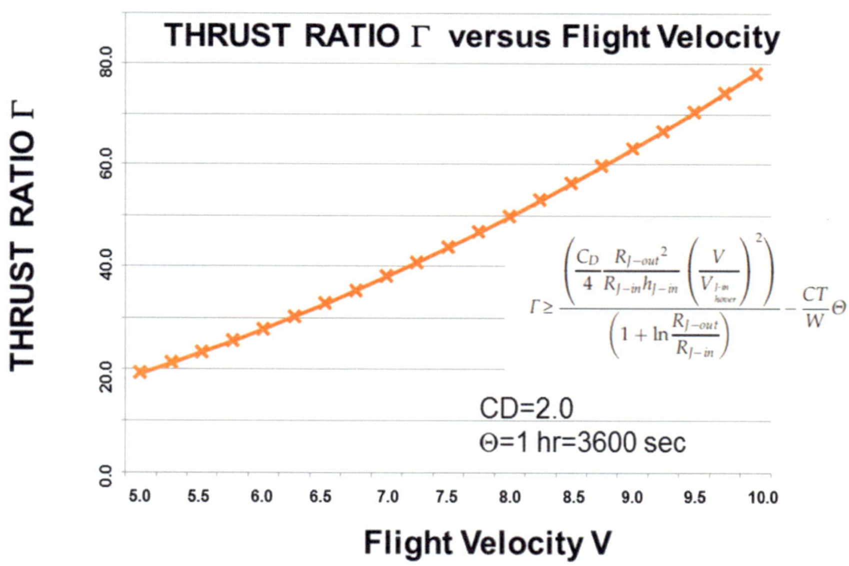

$$\Gamma \geq \frac{\left(\frac{C_D}{4}\frac{R_{J-out}^2}{R_{J-in}h_{J-in}}\left(\frac{V}{V_{J\text{-}in\,hover}}\right)^2\right)}{\left(1+\ln\frac{R_{J-out}}{R_{J-in}}\right)} - \frac{CT}{W}\Theta$$

Figure 24. Thrust Ratio Γ as a function of versus flight velocity for pure climbing, following Eq. (49) with equal sign, assuming C ≈ 2.0 and Specific Fuel Consumption C = 0.000139 and period = 1 h. The example is also worked out for the Coandă Jet inlet Radius $R_{J\text{-}in} = 0.02$ m, Coandă jet slot thickness $h_{J\text{-}in} = 0.01$ m, and Coandă jet inlet velocity $V_{J\text{-}in} =$ 5 m/sec.

results indicates the plausibility of the theoretical approach. Performance measures are derived to elaborate the significance of Coandă jet momentum coefficient and to assess the effectiveness of the Coandă MAV as an aerial robot platform.

Author details

Harijono Djojodihardjo

Address all correspondence to: harijono@djojodihardjo.com

The Institute for the Advancement of Aerospace Science and Technology, Jakarta, Indonesia

References

[1] Coandă H. Procédé et dispositif pour faire dévier une veine de fluide pénétrant dans un autre fluide. Brevet d'invention France. 1934

[2] Coandă H, France C. Device for Deflecting a Stream of Elastic Fluid Projected into an Elastic Fluid. US Patent Office, 2,052,869, 1936, ed.: Google Patents

[3] Gad-el-Hak M. Flow Control - Passive, Active and Reactive Flow Management. London: Cambridge University Press; 2000

[4] Liu Y, Sankar LN, Englar RJ, Ahuja KK. Numerical Simulations of the Steady and Unsteady Aerodynamic Characteristics of a Circulation Control Wing Airfoil AIAA Paper 2001-0704; 2001

[5] Kweder J, Panther C, Smith JE. Applications of circulation control, yesterday and today. International Journal of Engineering (IJE). 2011;4(5):411-429

[6] Englar RJ. Development of the A-6/Circulation Control Wing Flight Demonstrator Configuration. Defense Technical Information Center, Document. 1979

[7] Pugliese A, Englar RJ. Flight Testing the Circulation Control Wing. New York, N. Y: American Institute of Aeronautics and Astronautics, Aircraft Systems and Technology Meeting; 1979

[8] Mamou M, Khalid M. Steady and unsteady flow simulation of a combined jet flap and Coandă jet effects on a 2D airfoil aerodynamic performance. Revue des Energies Renouvelables CER'07 Oujda. 2007; 55-60

[9] Englar RJ, Hemmerly RA, Taylor DW, Moore WH, Seredinsky V, Walter W. Design of the circulation control wing STOL demonstrator aircraft. Journal of Aircraft. 1981;18:51-58

[10] Schroijen N, van Tooren M. MAV Propulsion System using the Coandă Effect. American Institute of Aeronautics and Astronautics. 45th AIAA/ASAME/SAE/ASEE. Exhibit. Joint Propulsion Conference, Denver, Colorado. 2009. pp. 1-10

[11] Mirkov N, Rasuo B. Numerical simulation of air jet attachment to convex walls and applications. International Congress of the Aeronautical Sciences; 2010. pp. 1-7

[12] Rumsey CL, Nishino T. Numerical study comparing RANS and LES approaches on a circulation control airfoil. International Journal of Heat and Fluid Flow. 2011;32:847-864

[13] Radespiel R, Burnazzi M. Fundamentals in Coandă Flap Design. Berlin Heidelberg: Springer-Verlag; 2011. https://www.researchgate.net/profile/Rolf_Radespiel/publication/283231590_Fundamentals_in_Coandă_Flap_Design/links/5698e9d108ae34f3cf207911/Fundamentals-in-Coandă-Flap-Design.pdf [Accessed: June 15, 2014]

[14] Djojodihardjo H. Progress and development of Coandă jet and vortex cell for aerodynamic surface circulation control – An overview. The SIJ Transactions on Advances in Space Research & Earth Exploration (ASREE). September-October 2013;1(1):32-42

[15] Djojodihardjo H, Thangarajah N. Research, development and recent patents on aerodynamic surface circulation control-A critical review. Recent Patents on Mechanical Engineering. 2014;7:1-37

[16] Tongchitpakdee C, Benjanirat S, Sankar LN. Numerical studies of the effects of active and passive circulation enhancement concepts on wind turbine performance. Journal of Solar Energy Engineering. 2006;128(4):31023089

[17] Tongchitpakdee C. Computational studies of the effects of active and passive circulation enhancement concepts on wind turbine performance [PhD Thesis]. Georgia Institute of Technology; 2007. https://www.scribd.com/document/38534330/Chanins-Thesis [Accessed 15 July 2015]

[18] Abdul-Hamid MF, Djojodihardjo H, Suzuki S, Mustapha F. Numerical assessment of Coandă effect as airfoil lift enhancer in wind-turbine configuration. In: Regional Conference on Mechanical and Aerospace Technology; 9-10 February 2010; Bali

[19] Djojodihardjo H, Abd Hamid MF, Basri S, Romli FI, Abdul Majid DLA. Numerical simulation and analysis of Coandă effect circulation control for wind-turbine application considerations. IIUM Engineering Journal, Special Issue, Mechanical Engineering. 2011. pp 19-42

[20] Djojodihardjo H, Abdul-Hamid MF, Jaafar A, Basri S, Romli FI, Mustapha F, Mohd-Rafie AS, Abdul-Majid. Computational study on the aerodynamic performance of wind turbine airfoil fitted with Coandă jet. Journal of Renewable Energy, Hindawi Publishing Co. 2013;2013:17, Article ID 839319, http://dx.doi.org/10.1155/2013/839319

[21] Erath B, Plesniak MW. The occurrence of the Coandă effect in pulsatile flow through static models of the human vocal folds. Journal of the Acoustical Society of America. August 2006;120(2):1000-1011

[22] Zheng X, Mittal R, Bielamowicz S. A computational study of asymmetric glottal jet deflection during phonation. Journal of the Acoustical Society of America. 2011;129(4):2133-2143

[23] Scheinherr A. Glottal motion and its impact on airflow and aerosol deposition in upper airways during human breathing [PhD Thesis]. Institut de Recherchesur les Phénomènes Hors Équilibre, Ecole Centrale du Marseille; 2015. https://hal.archives-ouvertes.fr/tel-01169649/document [Accessed 15 July 2015]

[24] Djojodihardjo H, Ahmed RI. CFD simulation of Coandă effect on the upper respiratory system. Journal of Medical Imaging and Health Informatic. October 2016;6(6):1526-1535(10)

[25] Prandtl L. Über Flüssigkeitsbewegung bei sehr kleiner Reibung Int. Math. Kongr. Heidelberg. Leipzig; 1904

[26] Frank A, McGrewy JS, Valentiz M, Levinex D, How JP. Hover, Transition, and Level Flight Control Michelson, C. Test and evaluation for fully autonomous micro air vehicles. ITEA Journal. 2008;29(4):1-12. Available from: http://citeseerx.ist. psu.edu/viewdoc/download?doi¼10.1.1.212.6553&rep¼rep1&type¼pdf [Accessed: April 30, 2015]

[27] Dumitrache A, Frunzulica F, Ionescu TC. Mathematical modelling and numerical investigations on the Coandă effect, Dumitrache et al. InTech; 2012. http://dx.doi.org/10.5772/50403 [Accessed: June 15, 2013]

[28] Constantinescu SG, Niculescu ML. Experimental and numerical research of lift force produced by Coandă effect. AIP Conference Proceedings. 2013;1558:152. DOI: 10.1063/1.4825443

[29] Mirkov N, Rasuo B. Maneuverability,of a UAV with Coandă Effect Based Lift Production. International Congress of the Aeronautical Sciences; 2012. pp. 1-6

[30] Ghozali D. Analysis of Coandă effect using computational-fluid-dynamic [Thesis]. Indonesia: Universitas Gajah Mada; 2013

[31] Djojodihardjo H, Ahmed RI. An analysis on the lift generation for Coandă micro air vehicles. In: IEEE, ICARES Conference Proceeding; Yogyakarta; 2014

[32] Djojodihardjo H, Ahmed RI, Abu-Thalib AR, Mohd-Rafie AS. Analytical and CFD visualization studies of Coandă MAV, In: Proceedings, The 13th Asian Symposium on Visualization; 22-26 June 2015; Novosibirsk, Russia. Russian Academy of Sciences and Kristianovich Institute of Theoretical and Applied Mechanics

[33] Ahmed RI, Djojodihardjo H, Abu-Talib AR, Abd-Hamid MF. Application of Coandă jet for generating lift of micro air vehicles - Preliminary design considerations. Applied Mechanics and Materials. 2014;629:139-144

[34] Ahmed RI, Djojodihardjo H, Abu-Talib AR, Mohd-Rafie AS. First principle analysis of Coandă micro air vehicle aerodynamic forces for preliminary sizing. Aircraft Engineering and Aerospace Technology. 2016;89(2):231-245. DOI: 10.1108/AEAT-03-2015-0080

[35] Djojodihardjo H, Ahmed RI. Basic Coandă MAV fluid dynamics and flight mechanics, Fifteenth Asian Congress of Fluid Mechanics (15ACFM) IOP Conference Series Journal of Physics: Conference Series 822, 012051, 2017

[36] Ahmed RI, Abu-Talib AR, Mohd-Rafie AS, Djojodihardjo H. Aerodynamics and flight mechanics of MAV based on Coandă effect. Aerospace Science and Technology. March 2017;62:136-147

[37] Davis Jr. WR, Kosicki B, Boroson DM, Kostishack DF. Micro air vehicles for optical surveillance. The Lincoln Laboratory Journal, MIT. 1996;9(2):197-214

[38] Poisson-Quinton Ph, Lepage L. Survey of French research on the control of boundary layer and circulation. In: Lachmann GV, editor. Boundary Layer and Flow Control: Its Principles and Application. Vol. 1. New York: Pergamon Press; 1961. pp. 21-73

[39] Van Kuik GAM. The relationship between loads and power of a rotor and an actuator disc. Journal of Physics: Conference Series. 2014;555:012101

[40] Bangura M, Melega M, Naldi R, Mahony R. Aerodynamics of Rotor Blades for Quadrotors, arXiv:1601.00733v1 [physics.flu-dyn] [Internet]. 5 January 2016. Available from: https://arxiv.org/pdf/1601.00733.pdf [Accessed: April15, 2017]

[41] Anonymous. Froudes' Momentum Theory: (Actuator Disk Theory). Available from: 17 http://www.public.iastate.edu /_aero442/unit2.pdf [Accessed: April 15, 2017]

[42] Anonymous. Froudes' Momentum Theory: (Actuator Disk Theory). Available from: http://www.public.iastate.edu /~aero442/unit2.pdf [Accessed: April 15, 2017]

[43] Betz A. Das Maximum der theoretisch moeglichen Ausnuetzung des Windes durch Windmotoren. Zeitschrift für das gesamte Turbinenwesen, 26, 307-309. (In German)

[44] Ahmed RI. Theoretical foundation and numerical analysis on Coandă Effect for micro-air vehicle [PhD Thesis]. Universiti Putra Malaysia; 2016

[45] Anonymous. ANSYS® ICEM CFD Tutorial Manual. Available from: 148.204.81.206/Ansys/150/ANSYS%20ICEM%20 CFD%20 Tutorial%20Manual.pdf [Accessed: January 15, 2011]

Chapter 3

Innovative Propulsion Systems and CFD Simulation for Fixed Wings UAVs

Esteban Valencia and Victor Hidalgo

Additional information is available at the end of the chapter

http://dx.doi.org/10.5772/intechopen.70273

Abstract

Nowadays, mobile applications demand, in large extent, an improvement in the overall efficiency of systems, in order to diversify the number of applications. For unmanned aerial vehicles (UAVs), an enhancement in their performance translates into larger payloads and range. These factors encourage the search for novel propulsion architectures, which present high synergy with the airframe and remaining components and subsystems, to enable a better UAV performance. In this context, technologies broadly examined are distributed propulsion (DP), thrust split (TS), and boundary layer ingestion (BLI), which have shown potential opportunities to achieve ambitious performance targets (ACARE 2020, NASA N+3). The present work briefly describes these technologies and shows preliminary results for a conceptual propulsion configuration using a set number of propulsors. Furthermore, the simulation process for a blended wing body (BWB) airframe using computational fluid dynamics (CFD) OpenFOAM software is described. The latter is examined due to its advantages in terms of versatility and cost, compared with licensed CFD software. This work does not intend to give a broad explanation of each of the topics, but rather to give an insight into the state of the art in modeling of distributed propulsion systems and CFD simulation using open-source software implemented in UAVs.

Keywords: unmanned aerial vehicles, CFD, distributed propulsion, boundary layer ingestion, blended wing body, OpenFOAM

1. Distributed propulsion

Distributed propulsion has been studied as a way of improving propulsive efficiency by increasing the bypass ratio of the turbofan engines, without any design constraints of large engines' fan radius. Distributed propulsion has been explored since the 1960s, for commercial aviation; however, the design complexities involving energy transmission, maintenance, and fluctuant oil price during the past decades have limited its research. Recently, the growing environmental awareness has again motivated initiatives for the search of more efficient

INTECH
open science | open minds

© 2017 The Author(s). Licensee InTech. This chapter is distributed under the terms of the Creative Commons Attribution License (http://creativecommons.org/licenses/by/3.0), which permits unrestricted use, distribution, and reproduction in any medium, provided the original work is properly cited.

propulsion systems. Numerous renowned institutions [1–3] have taken the leap to study concepts with distributed propulsion and observed an achievable 8% benefit in terms of thrust specific fuel consumption (TSFC) compared with today's aircraft [2]. However, for civil aviation, the implementation of this technology is still a challenge, due to mechanical design constraints and safety issues. For unmanned platforms, this technology has been implemented successfully on multicopters [4], where multiple rotors are used for propulsion and flight control. The latter application has arisen as a consequence of suitable and reliable electrical power, control, and transmission systems, which enable to control pitch, torque, and rotational speed, to describe the flight envelope. Although distributed propulsion has been well examined in multicopters, unmanned aerial platforms, using fixed wing configurations, are not well documented. For fixed wings, the distributed propulsors could be arranged over the airframe [5], and they could present boundary layer ingestion, which could bring 7% reduction for civil aviation concepts in TSFC compared with today's aircraft [6]. Since highly integrated propulsion configurations have a high synergy with BWB airframes, the latter has been selected for the assessment of the propulsion configurations, and hence to set the thrust required at cruise conditions, the aerodynamic characteristics are based on a BWB airframe. A plot of the BWB airframe selected for this work is shown in **Figure 1**, and as observed, its selection obeys the multiple advantages for DP and BLI, such as (i) its uniform planform area in the spanwise direction, which enables the allocation of an array of propulsors; (ii) the large airframe space to embed the propulsors at the trailing edge of the airframe, which is beneficial when BLI is implemented, as more airframe drag is ingested. In the CFD simulation presented in this work, the BWB configuration is tested using the OpenFOAM open-source software. The aerodynamic

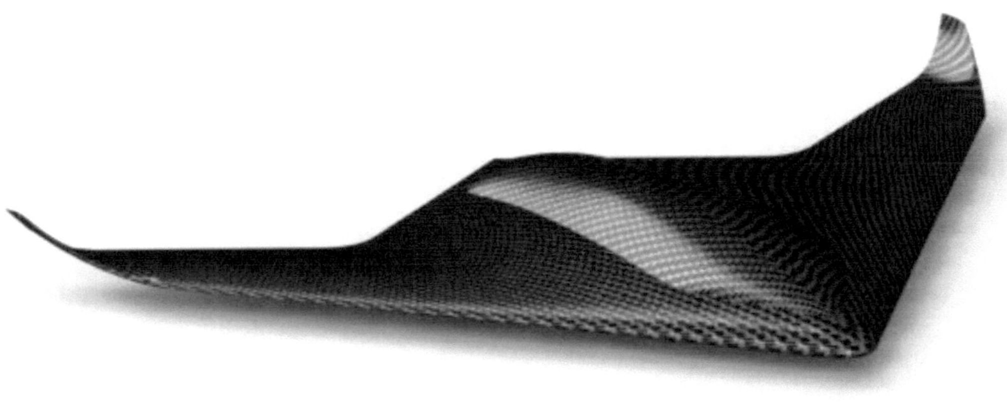

Figure 1. BWB model selected [7].

C_{DO} (zero lift, drag coefficient)	C_L (Lift coefficient)	e (wing span efficiency)	AR (aspect ratio)	Flight velocity	Wing span
0.015	0.1	0.72	5.9	20 (m/s)	3 (m)

Table 1. Airframe and aerodynamic data at cruise conditions [8].

characteristics of the BWB airframe are described in **Table 1**, which correspond to the UAV case study. Since agriculture is one of the main applications for this sort of aerial platforms, the operating conditions are selected based on UAVs used in precision agriculture [5].

In order to clarify the structure of the distributed propulsion configuration, an illustrative diagram for the electrical propulsion system utilized in this work is shown in **Figure 2**. As observed in this figure, the electrical and control system units take the place of the gas turbine in common turbo-electric distributed propulsion arrangements [9], and hence they supply the electrical power to the propulsor's electrical motor. For the sake of clarity, in **Figure 2**, only one propulsor (NF = 1) is depicted; nevertheless, this is a design space variable for the UAV concept.

For the case of study previously mentioned and using the methodology developed in Valencia et al. [5], the power consumed by the distributed propulsion arrangement using an electrical power system is shown in **Figure 3**. The figure shows the power requirement per fan to achieve the set thrust at cruise condition for four propulsion configurations. In this case, each configuration uses a different number of propulsors (NF). To calculate the mass flow through the propulsors, an inner control volume approach is selected [10]. Through this approach, the propulsor mass flow can be defined based on the set thrust for cruise condition, which is calculated as a function of the aircraft drag. Eq. (1) is used to calculate the propulsor's mass flow. The latter variable enables to calculate the fan diameter through mass conservation and the fan power based on the isentropic thermodynamic relations for incompressible flow. This process is further explained in [3]

$$F_N = \dot{m}\left(V_j - V_0\right) \tag{1}$$

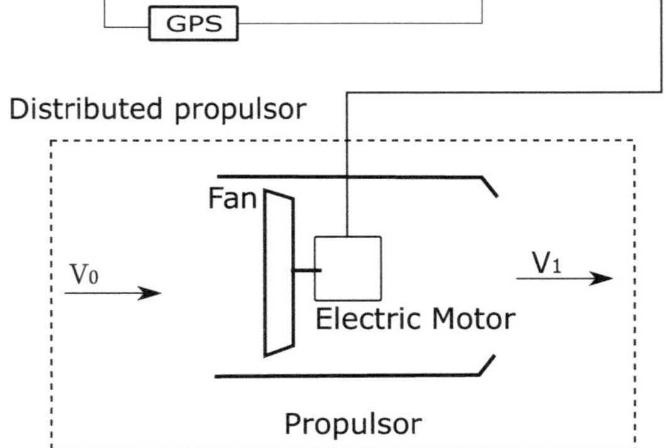

Figure 2. Distributed propulsion array for an electric-powered UAV.

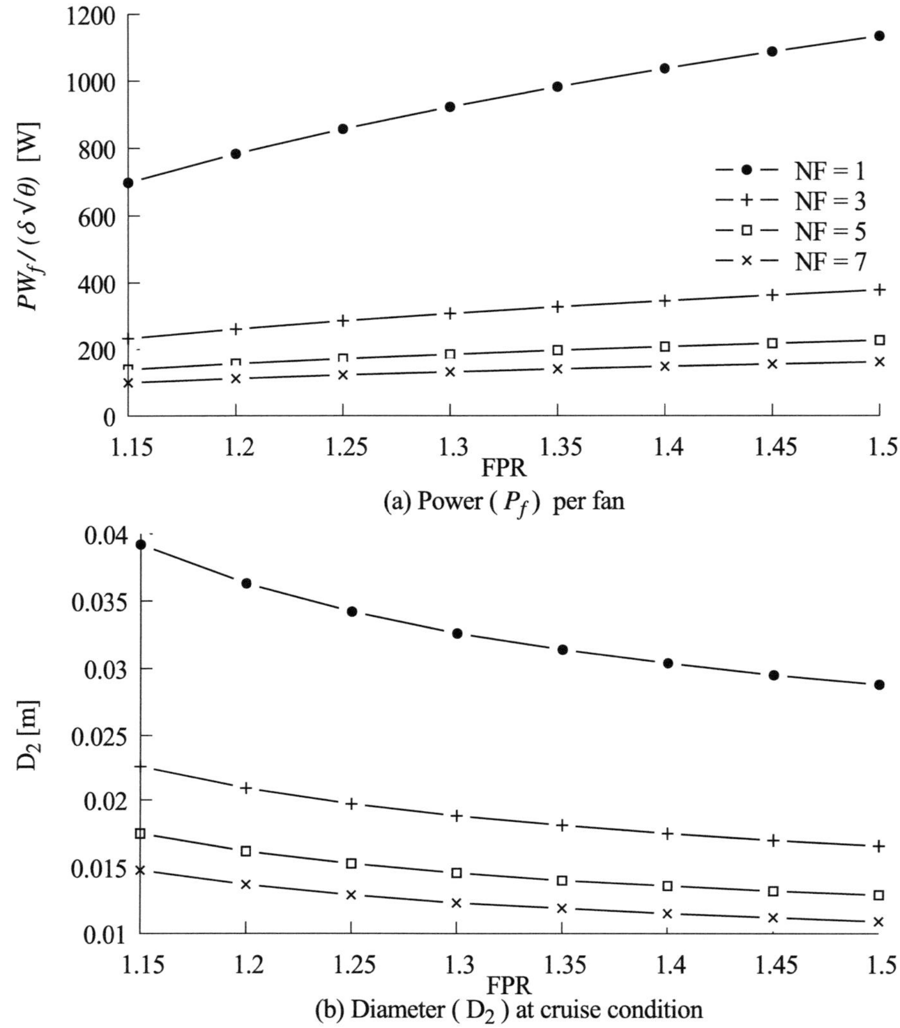

Figure 3. Power required per fan and fan diameter at cruise condition.

In **Figure 3(a)**, the total power remains the same, as it is assumed that the aircraft drag remains constant (this assumption, however, will depend on the distributed propulsion array); furthermore, for the calculation of the propulsor power, it is assumed that the flow entering the distributed propulsors presents the same velocity for all the propulsors. In **Figure 3(a)**, it is observed how the power required per fan reduces as more fans are implemented in the distributed arrangement and additionally shows the benefit of working with low fan pressure ratios. In this case, the intake pressure losses assumed are 1%, which is low and hence they do not restrict to operate at low fan pressure ratios. This, however, will highly depend on the duct design. It can also be observed from the figure that the total power of the distributed arrangement remains constant for all the configurations. In **Figure 3(b)**, the fan diameter for each arrangement is plotted and, as expected, a large number of propulsors will reduce their size, due to the lower mass flow per propulsor. Also, it is important to note that insofar as the pressure ratio is increased, the fan diameter decreases; this is attributed to the larger energy that is delivered by the fan to produce the thrust set for cruise condition. To summarize, the benefits that distributed propulsion brings are (i) enhancement in the distribution of loads along the spanwise of the airframe, due to the reduction in size and therefore weight of the propulsors; (ii) suitability of small fans embedded into the airframe to improve its aerodynamic performance.

In the case of using turbo-electric distributed propulsion, where the energy source is a gas turbine/turbofan, the propulsive efficiency will improve, due to the increment of energy transferred to the low momentum flow (cold flow), and hence, the reduction of velocity in the high speed flow [11]. The latter, also, will contribute to a reduction in noise [12].

BLI configurations having a distributed propulsor array over the airframe sucking the boundary layer induce certain challenges in terms of boundary layer treatment, stability issues, aerodynamic integration effects propulsors-airframe, among others. It is usually a problem to deal with the boundary layer around the airframe, as highly coupled configurations ingest the boundary layer, and hence, the propulsors operate under combined radial and circumferential distortion patterns, which detriments to large extent fan performance [2, 5]. **Figure 4** shows the total pressure patterns at different radial and circumferential positions for a clean fan and for the BLI case. As observed, the change in flow properties of each rotational cycle would affect the fan performance, since blades are designed only for radial distortion, where a set flow conditions is expected at the design point. There are novel techniques to deal with boundary layer ingestion and take advantage of a reduced momentum drag for the propulsors. These alternatives either enhance control of the incoming flow or shift the blade position based on the

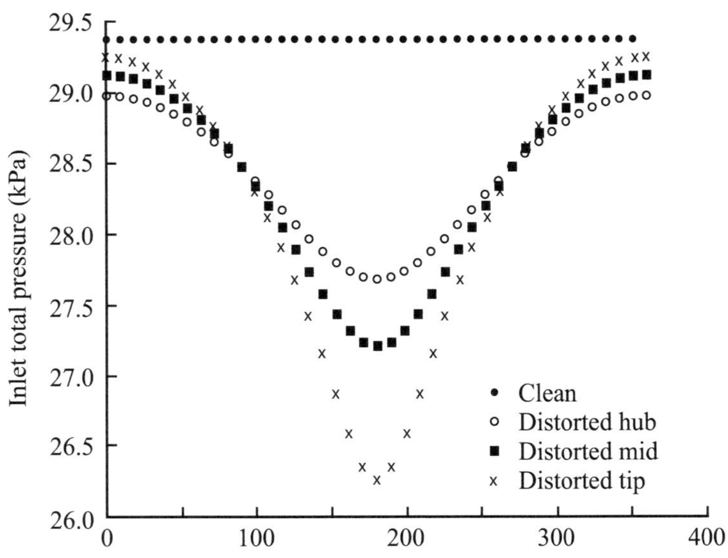

Figure 4. Inlet total pressure for distorted BLI case [15].

flow conditions [13, 14]. The discussion of these alternatives is beyond the scope of the present work.

1.1. Thrust split

The concept of thrust split has been well explored in many different manned aerial concepts, such as the N3-X [1, 6] and Cranfield [3, 16, 17]. This technology, for manned concepts, which presents distributed propulsion, was observed as a way of enhancing propulsive efficiency and fuel consumption, while reducing the high transmission losses (electrical transmission systems) and integration aerodynamic effects between propulsion and airframe [10]. The mentioned benefits are achieved in civil aviation, by reducing the amount of power required by the distributed propulsion arrangement, and hence decreasing the amount of losses and increasing weight, which comes from the cryocooling system of the high-temperature superconducting (HTS) electrical propulsion [18]. Regarding BLI systems, it was observed that ingesting a larger share of boundary layer freestream determines the performance benefits from BLI [16] and thus, it is better to have smaller propulsors, which ingest only the boundary layer. The improvement in propulsive efficiency and energy consumption comes from the lower momentum drag entering the distributed propulsor array. Furthermore, the latter aspect can have other positive effects with regard to intake pressure losses and distributed array allocation. The first aspect is due to the reduced mass flow required by the propulsor array operating with BLI, and hence, less wetted area at the intake. The second aspect is related to the small size of the propulsors and hence less geometrical constraints for their allocation in the distributed propulsor unit. Thrust split (T_S) is defined by Eq. (2).

$$T_S = \frac{F_{DP}}{F_N} \tag{2}$$

where F_{DP} is the thrust delivered by the propulsor array and F_N is the net intrinsic net thrust [3].

In case of battery-powered aerial platforms, which have the configuration presented in **Figure 2**, thrust split is an extra design space variable for the configuration of distributed propulsion systems with BLI. This variable could help heterogeneous fan distributed propulsors arrangements, where geometric, aerodynamic, and structural constraints can be satisfied. **Figure 5** shows a conceptual configuration of the UAV described in **Table 1**, with a thrust split of 50% and 75%. In this case, the main engine is located at the centerline to reduce stability problems in case of one engine is put off. It is observed from the figures how the heterogeneous configurations for UAVs could help to better the distribution of loads (propulsors) within the airframe, versatility in configurations and better aerodynamic design (smaller propulsors embedded in the airframe). In **Figure 5**, the main fan power corresponds to the airframe centerline fan shown in **Figure 6**.

Illustrative configurations for the three thrust split configurations presented in **Figure 5** are shown in **Figure 6**. As observed in the figure, insofar as thrust split is reduced, the size of the main propulsor increases, which is attributed to the large share of thrust that needs to be delivered by the main propulsor to keep the thrust at cruise condition.

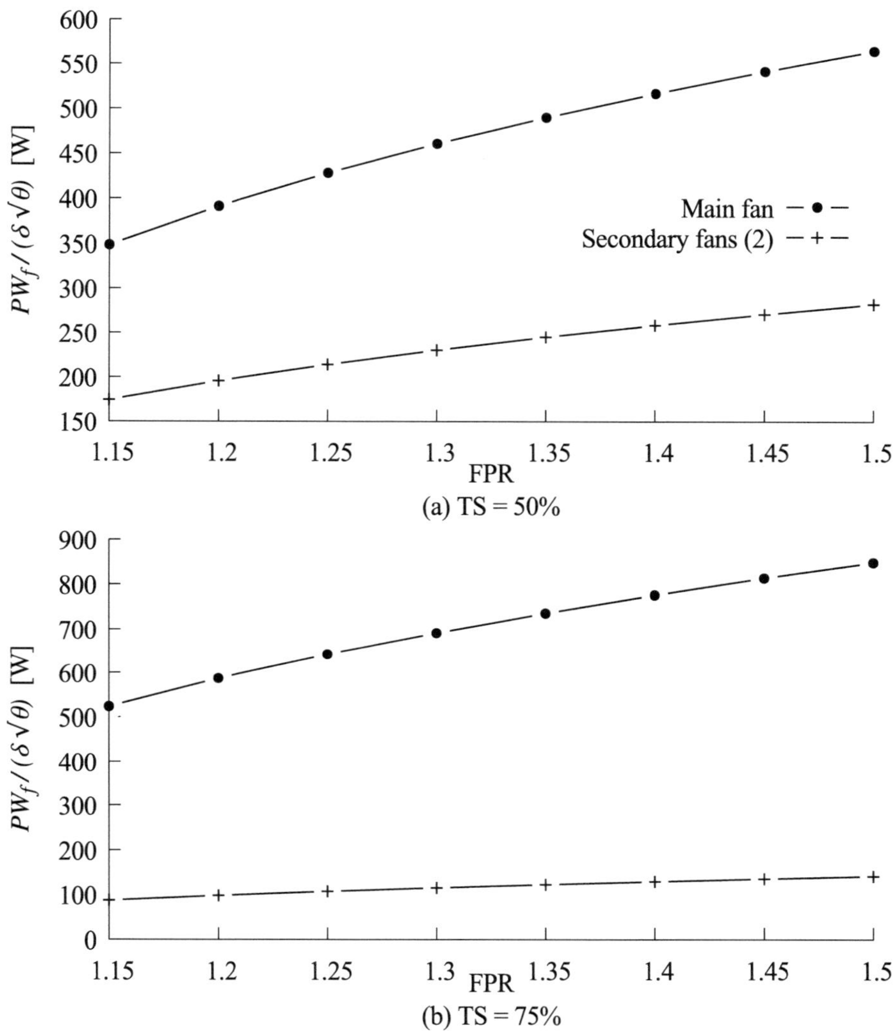

Figure 5. Propulsion system configurations for 50 and 75% thrust split.

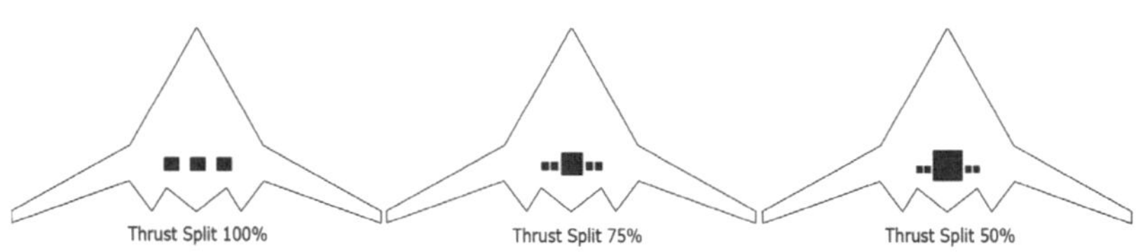

Figure 6. Illustrative configurations for different thrust splits.

In unmanned aerial vehicles, which use distributed propulsion, gas turbines or internal combustion engines work as the main power source. These previously mentioned benefits need to be verified and adapted for the different power settings that UAVs required [5]. First, most of unmanned systems do not employ turbofans, due to increased cost and complexity of these systems; therefore, the benefit in propulsive efficiency accrued from using large by pass ratio turbofans in high thrust split configurations will not be present. Nevertheless, UAVs that have gas turbines and distributed propulsion may have structural issues that come from the load distribution within the airframe. This latter aspect could be solved by introducing thrust split as main power source, and the propulsion system (main and distributed) could be sized based on a thrust schedule where the main engine and the propulsor array can cooperate with a share of the thrust required at each flight condition during the flight envelope looking for an optimum in terms of weight and overall performance.

2. CFD simulation

The fixed-wing UAVs can be examined using CFD simulation to reduce cost in investigation [19, 20]. For that purpose, OpenFOAM have been used to carry out a numerical simulation for a general shape of UAVs' wing. OpenFOAM means Open Field Operation and Manipulation and it is not only a software but also a library of C++ solvers for CFD simulations. The main code is free and open-source software, which allows to modify and to implement equations and functions without any commercial restriction [20]. In this context, OpenFOAM is an important tool to carry out numerical simulations for CFD.

2.1. Description of the mathematical model

The turbulence model of $SST\ k-\omega$ has been selected to carry out the numerical simulation to capture the effects of the boundary layer [21] over the airframe, which in this case presents geometrical complexity. The adverse pressure gradient and the separating flow can be captured fairly well with considerations of $SST\ k-\omega$ properties in regions far away from walls [21]. Moreover, a general expression of the turbulence model is indicated in the filtered Eq. (3).

$$\rho \frac{\partial \overline{\phi}}{\partial t} + \rho \overline{u}_j \frac{\partial \overline{\phi}}{\partial x_j} - \frac{\partial}{\partial x_j}\left[\Gamma_{\phi,\text{eff}} \frac{\partial \overline{\phi}}{\partial x_j} \right] = S_\phi \tag{3}$$

where ϕ represents variables, $\Gamma_{\phi,\text{eff}}$ represents the effective diffusion coefficient, and S_ϕ represents the source term of the equation, which is according to [21].

The OpenFOAM solver called simpleFoam has been used for solving the steady-state Reynolds-averaged Navier-Stokes equations with the $SST\ k-\omega$ turbulence model [22]. For that, the coupling between velocity and pressure is treated using the SIMPLE method [23].

2.2. CFD domain

The selected geometry of UAVs for numerical simulations has been based on the design of the BWB airframe [24, 25] shown in **Figure 1**. The analyzed 3D model is depicted in **Figure 7**.

The main dimensions of the domain are indicated in **Figure 8**, which was based on previous work focused on drag estimations using CFD [26]. The largest dimension is 15 times the chord

http://dx.doi.org/10.5772/intechopen.70273

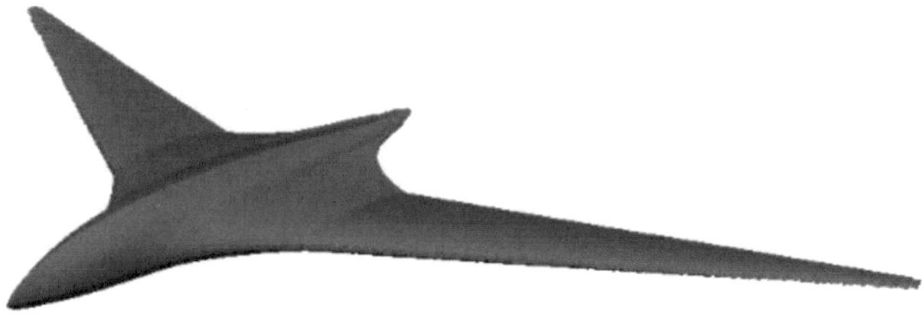

Figure 7. 3D model of BWB Mark 2 [24, 25].

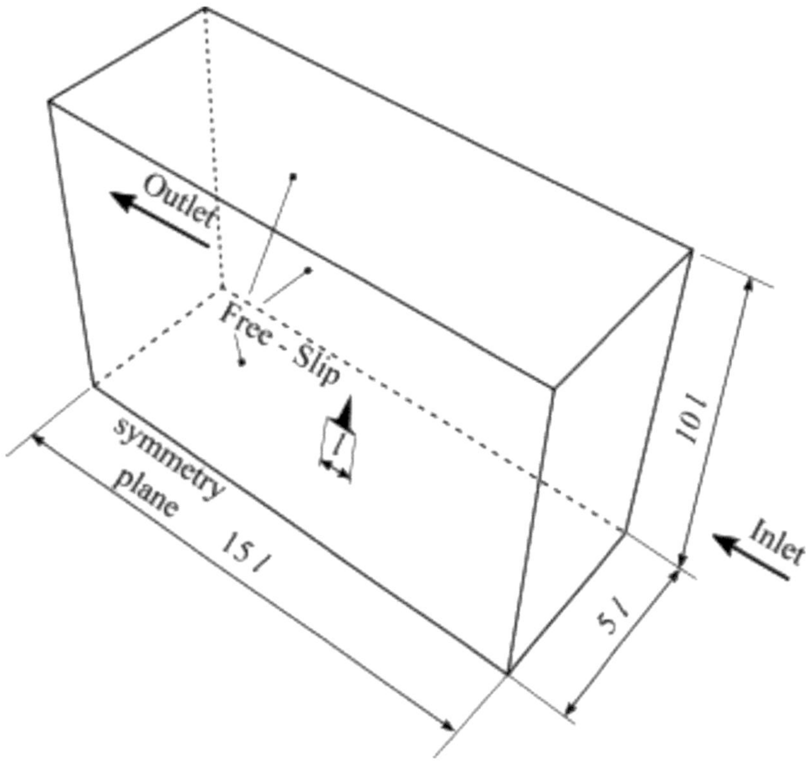

Figure 8. Main computational domain.

length, *l*, in the axial direction and the shortest length is 5 times *l* in the spanwise direction. The aforementioned configuration was used for the CFD simulation.

2.3. Mesh generation and boundary conditions

The mesh in **Figure 9** was obtained using the OpenFOAM tool snappyHexMesh [27] with 1.3 million cells. Based on previous research [19, 28], the mesh has been divided in two zones, refined and no-refined, to capture special characteristics of the phenomenon and fluid development, respectively. Furthermore, the refined zone close to the wing walls presents a special

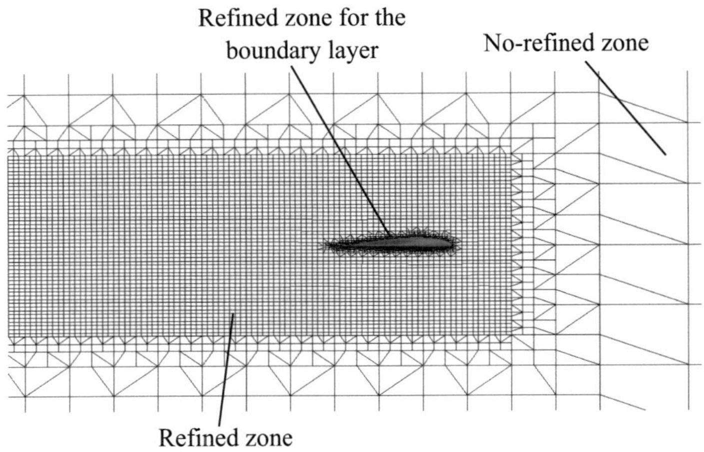

Figure 9. Mesh around the BWB Mark 2 model.

U_{in}	p_{out}	ρ_∞	μ_∞
[m/s]	[m²/s²]	[kg/m³]	[kg/(ms)]
50	0	0.5895	1.561×10^5

Table 2. Boundary conditions.

treatment to improve the simulation of the boundary layer, which has a y^+ equal to 50 according to the requirements of the $SST\ k - \omega$ model [19, 27].

The boundary conditions are based on previous studies for BWB [29] and indicated in **Table 2**, where U_{in} is the inlet velocity, p_{out} is the static pressure in the outlet, ρ_∞ is the estimated density of the air, and μ_∞ is the dynamic viscosity.

2.4. Results and discussion

Figure 10 shows that the numerical simulation converge after 200 iterations for the pressure and components of the velocity, U, in the respective axes; hence, 300 iterations can be acceptable for steady-state results.

According to **Figure 11(a)**, the pressure coefficient (C_p) distribution of the BWB Mark 2 model changes from the leading edge to the trailing edge. The upper part indicated in **Figure 11(b)** shows the C_p decreasing downstream and the lowest C_p was found close to the quarter chord of wings, maintaining the trends as expected for BWB airframes [8, 26]. Downstream, the minimum C_p increases in the direction of the trailing edge. The bottom view shows similar C_p distribution to the upper surface with higher values to give the lifting characteristics of this component.

Furthermore, the C_L and C_D were calculated in OpenFOAM using the forces library called libforces. The estimated values are 0.216 and 0.014 for C_L and C_D, respectively, which matched fairly well the values obtained by Cisneros [29] using XLFR5. Since the aim of the present work

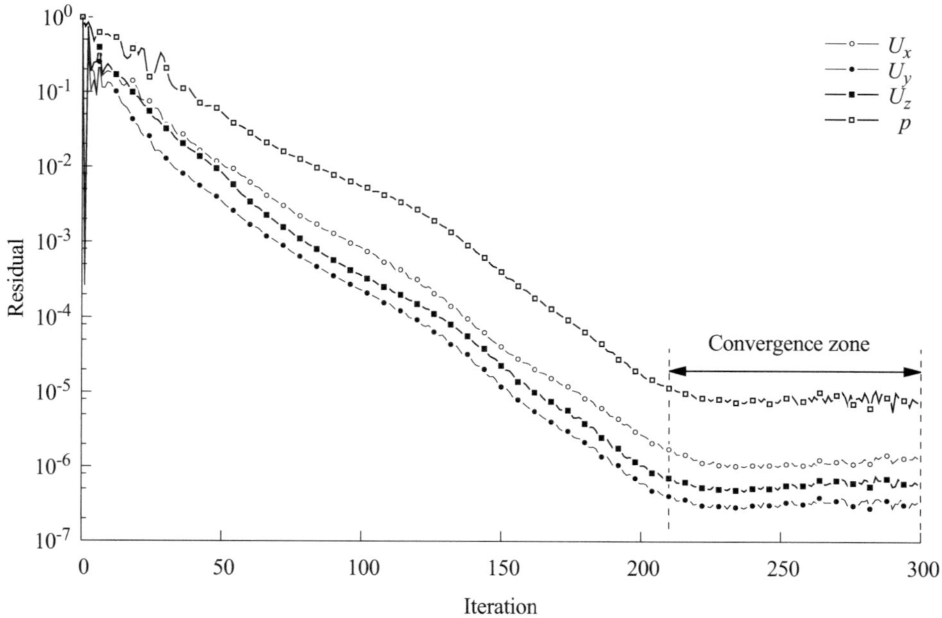

Figure 10. General residuals of the BWB Mark 2 simulation.

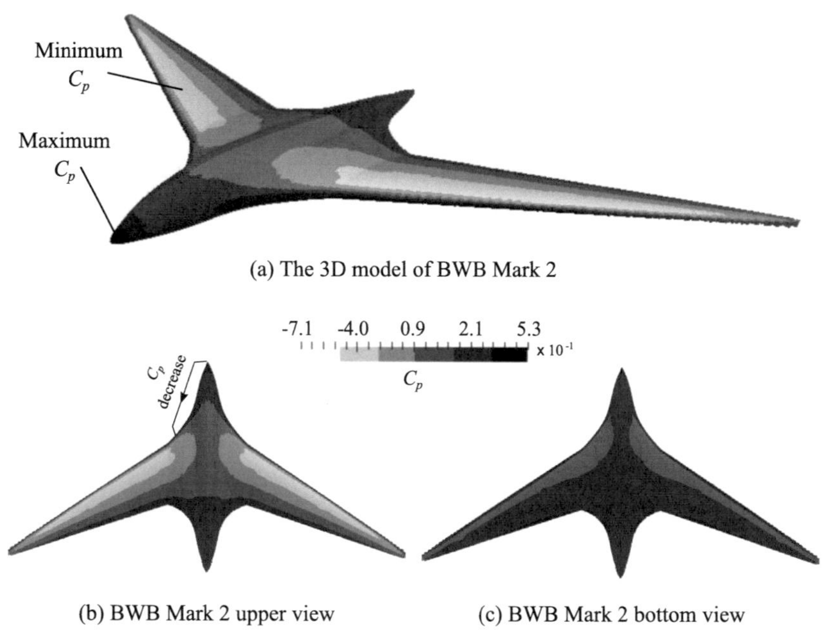

Figure 11. Pressure distribution around the BWB Mark 2 model.

is to give an insight on high-fidelity aerodynamic assessment using open-source software and demonstrate the suitability of OpenFOAM, the simulation for other flight conditions is beyond the scope of the present work.

3. Summary

This study assessed the suitability of alternative propulsion configuration for UAVs, using a blended wing body (BWB) airframe with distributed propulsion and thrust split. Since the airframe design influences in the propulsion system, a prototype configuration was tested for the aerodynamic performance assessment through OpenFOAM, which is a library of C++ solvers for CFD simulation under guidelines of the free and open-source software. In this context, it was demonstrated that distributed propulsion, together with thrust split, enables structural, aerodynamic, and performance benefits. The aforementioned was achieved through a better load distribution along the airframe spanwise and reduction of propulsors' size to allow embedding them to attain a low drag propulsion array. Furthermore, these latter configurations open the window for BLI installations, which bring important benefits in terms of propulsive efficiency and drag reduction, but represent challenges from the aerodynamic integration perspective.

Finally, the research demonstrated the suitability of using OpenFOAM as CFD platform for aerodynamic performance assessment, which presents significant advantages in terms of (i) freedom to adapt the main code to our needs and (ii) no license requirement, which is important to reduce simulation costs.

Acknowledgements

The authors gratefully acknowledge the financial support provided by Escuela Politécnica Nacional for the development of the project PIJ 15-11, PIS 16-20 and PIMI 15-03.

Abbreviations and nomenclature

UAV_S	Unmanned Aerial Vehicle System
BLI	Boundary Layer Ingestion
DP	Distributed Propulsion
BWB	Blended Wing Body
TSFC	Thrust Specific Fuel Consumption
FPR	Fan Pressure Ratio
CFD	Computational Fluid Dynamics

HTS	High Temperature Superconducting
SST	Shear Stress Transport
LP	Low Pressure
HP	High Pressure
C_L	Lift Coefficient
C_D	Drag Coefficient
C_p	Pressure Coefficient
C_{Do}	Zero Lift Drag Coefficient
$\dot{m}\cdot$	Mass Flow
AR	Aspect Ratio
e	Wing Span Efficiency
P_f	Power Per Fan
θ	Referred Temperature
δ	Referred Pressure
D_2	Fan Diameter
N_f	Number of Fans
T_S	Thrust Split
F_{DP}	Thrust Delivered by the Propulsor Array
F_N	Net Thrust
V_0	Velocity at Propulsor Intake
V_1	Velocity at Propulsor Outlet
U_{in}	Air Velocity at Inlet for CFD Simulation
p_{out}	Air Gaussian Pressure at Outlet
ρ_∞	Air Density
μ_∞	Air Viscosity
η_f	Fan Efficiency
ϕ	Variables
$\Gamma_{\phi,eff}$	Effective Diffusion Coefficient
S_ϕ	Source Term

Author details

Esteban Valencia* and Victor Hidalgo

*Address all correspondence to: esteban.valencia@epn.edu.ec

Departamento de Ingeniería Mecánica, Escuela Politécnica Nacional, Quito, Ecuador

References

[1] Felder J, Kim H, Brown G. Turboelectric distributed propulsion engine cycle analysis for hybrid-wing-body aircraft. In: 47th AIAA Aerospace Sciences Meeting Including the New Horizons Forum and Aerospace Exposition; 2009. p. 1132

[2] Plas A. Performance of a boundary layer ingesting propulsion system [MSc. thesis], Massachusetts Institute of Technology, [MSc. thesis], Dept. of Aeronautics and Astronautics, Massachusetts, USA, 2006

[3] Valencia EA, Nalianda D, Laskaridis P, Singh R. Methodology to assess the performance of an aircraft concept with distributed propulsion and boundary layer ingestion using a parametric approach. Proceedings of the Institution of Mechanical Engineers, Part G: Journal of Aerospace Engineering. 2015;229(4):682-693

[4] Gundlach J. Designing Unmanned Aircraft Systems: A Comprehensive Approach. American Institute of Aeronautics and Astronautics, Inc.1801 Alexander Bell Drive, Reston, Virginia, 2012, 20191-4344

[5] Valencia E, Hidalgo V, Cisneros J. Design point analysis of a distributed propulsion system with boundary layer ingestion implemented in UAV's for agriculture in the Andean region. In: 52nd AIAA/SAE/ASEE Joint Propulsion Conference, Salt Lake City, Utah, 2016

[6] Felder J, Kim H, Brown G, Kummer J. An examination of the effect of boundary layer ingestion on turboelectric distributed propulsion systems. In: 49th AIAA Aerospace Sciences Meeting Including the New Horizons Forum and Aerospace Exposition, 4-7 January 2011, Orlando, Florida, 2011. p. 300

[7] Thompson D, Feys J, Filewich M, Abdel-Magid S, Dalli D, Goto F. The design and construction of a blended wing body UAV. In: 49th AIAA Aerospace Sciences Meeting Including the New Horizons Forum and Aerospace Exposition, 4-7 January 2011, Orlando, Florida, 2011. p. 841

[8] Lehmkuehler K, Wong K, Verstraete D. Design and test of a UAV blended wing body configuration. In: Proceedings of the 28th Congress of the International Council of the Aeronautical Sciences, Brisbane, Australia; September 2012; 2012. pp. 23-28

[9] Liu C, Valencia E, Teng J. Design point analysis of the turbofan-driven turboelectric distributed propulsion system with boundary layer ingestion. Proceedings of the Institution of Mechanical Engineers, Part G: Journal of Aerospace Engineering. 2016;230(6):1139-1149

[10] Rodriguez DL. Multidisciplinary optimization method for designing boundary-layer-ingesting inlets. Journal of Aircraft. 2009;46(3):883-894

[11] Kirner R. An Investigation into the Benefits of Distributed Propulsion on Advanced Aircraft Configuration, Dept. of Power and Propulsion, Cranfield University, Cranfiled, UK, 2013

[12] de la Rosa Blanco E, Hall C, Crichton D. Challenges in the silent aircraft engine design. In: 45th AIAA Aerospace Sciences Meeting and Exhibit, Jan 8-11, 2007, Reno, Nevada; 2007. p. 454

[13] Wang J-J, Choi K-S, Feng L-H, Jukes TN, Whalley RD. Recent developments in DBD plasma flow control. Progress in Aerospace Sciences. 2013;62:52-78

[14] Gorton S, Owens L, Jenkins L, Allan B, Schuster E. Active flow control on a boundary-layer-ingesting inlet. In: 42nd AIAA Aerospace Sciences Meeting and Exhibit, 5-8 Jan. 2004; Reno, Nevada; United States; 2004. p. 1203

[15] Valencia E, Hidalgo V, Nalianda D, Panagiotis L, Singh R. Discretized Miller approach to assess effects on boundary layer ingestion induced distortion. Chinese Journal of Aeronautics. 2017;30(1):235-248

[16] Valencia EA, Hidalgo V, Panagiotis L, Nalianda D, Singh R, Liu C. Design Point Analysis of an Hybrid Fuel Cell Gas Turbine Cycle for Advanced Distributed Propulsion Systems. American Institute of Aeronautics and Astronautics, In: 51st AIAA/SAE/ASEE Joint Propulsion Conference Orlando, Florida, 2015

[17] Liu C. Turboelectric distributed propulsion system modelling [PhD thesis]. Dept. of Power and Propulsion, Cranfield University, Cranfiled, UK; 2013

[18] Stroetmann C. GmbH, Style of Speed - Air - Plane - Hybrid - BWB-1/White Eagle, http://www.styleofspeed.com/air/plane/hybrid/bwb-1/index.htm

[19] Hidalgo V, Luo X-w, Escaler X, Ji B, Aguinaga A. Implicit large eddy simulation of unsteady cloud cavitation around a plane-convex hydrofoil. Journal of Hydrodynamics, Series B. 2015;27(6):815-823

[20] Hidalgo V. Numerical study on unsteady cavitating flow and erosion based on homogeneous mixture assumption [PhD thesis]. Beijing: Tsinghua University; 2016

[21] Zhang Z, Zhai ZJ, Zhang W, Chen QY. Evaluation of various turbulence models in predicting airflow and turbulence in enclosed environments by CFD: Part 2–comparison with experimental data from literature. HVAC&R Research. Nov. 2007;13(6):871-886

[22] Page M, Beaudoin M, Giroux A-M. Steady-state capabilities for hydroturbines with OpenFOAM. International Journal of Fluid Machinery and Systems. 2011;4(1):161-171

[23] Ferziger J.H. and Peric M., Computational Methods for Fluid Dynamics, Springer-Verlag Berlin Heidelberg, Nuernberg, Germany, 2002, doi: 10.1007/978-3-642-56026-2

[24] Qin N, Vavalle A, Le Moigne A, Laban M, Hackett K, Weinerfelt P. Aerodynamic considerations of blended wing body aircraft. Progress in Aerospace Sciences. 2004;40(6):321-343

[25] Kuntawala NB. Aerodynamic shape optimization of a blended-wing-body aircraft configuration [PhD thesis]. Institute for Aerospace Studies, University of Toronto, Ontario, Canada, 2011

[26] Hardie S. Drag Estimations on Experimental Aircraft Using CFD. Mälardalen University, Department of Mathematics and Physics, Sweden, 2007

[27] Greenshields C. OpenFOAM - Programmer's Guide, version 2.4.0 ed. OpenFOAM Foundation Ltd.; May 2015

[28] Krajnovi S, Lrusson R, Basara B. Superiority of PANS compared to LES in predicting a rudimentary landing gear flow with affordable meshes. International Journal of Heat and Fluid Flow. Oct. 2012;37:109-122

[29] Cisneros Gallegos JA. Design and Simulation of a Blended Wing Body (BWB) Airframe for an Unmanned Aerial Vehicle (UAV) Using Computational Fluid Dynamics (CFD); 2015

Chapter 4

Comparison between Semiempirical and Computational Techniques in the Prediction of Aerodynamic Performance of the Rotor of a Quadcopter

Andres Mauricio Pérez Gordillo

Additional information is available at the end of the chapter

http://dx.doi.org/10.5772/intechopen.69730

Abstract

Unmanned aerial vehicle (UAV) is a growing technology used in different industries, and the main platform used for the UAVs is the quadcopter. The rotor of a quadcopter typically operates at low to moderate Reynolds number, so that the aerodynamics and an early prediction of the performance of the propellers are important in the design of the quadcopter. In the present chapter, the performance of a commercial propeller used in quadcopters is analyzed with three different techniques: momentum theory, blade element theory, and computational fluid dynamics. By applying the momentum and blade element theory, it was possible to estimate the thrust generated for a propeller in hover. A computational model based on computational fluid dynamics (CFD) was implemented and used to simulate a propeller in hover; the model predicts the wake and the thrust of the propeller as well. The results of the theory and computational approximations were compared with experimental measurements of flying tests.

Keywords: momentum theory, blade element theory, CFD, lift coefficient, figure of merit, thrust

1. Introduction

The increasing use of the unmanned aerial vehicles (UAVs) in different industries has led to the development of the common platforms used in the design of UAVs. The preferable of these platforms is the quadrotor due to its advantages such as an easier control in flight for drone

INTECH
open science | open minds

© 2017 The Author(s). Licensee InTech. This chapter is distributed under the terms of the Creative Commons Attribution License (http://creativecommons.org/licenses/by/3.0), which permits unrestricted use, distribution, and reproduction in any medium, provided the original work is properly cited.

operators, a rapid takeoff, and maneuverability. In the design of a quadcopter, it is necessary to estimate the performance of the propellers used in the vehicle. This chapter presents two of the simplest and more used theories to achieve a preliminary approximation to the performance of the selected propeller; also, a CFD model is used to study the aerodynamics of a commercial propeller used in a quadcopter of the Colombian company ADVECTOR.

Momentum theory is the simplest theory to study a propeller. A propeller produces thrust by the acceleration of the column of air that passes through the rotor plane, and with the use of the laws of conservation of mass, momentum, and energy, it is possible to obtain a relation of the induced velocity from the propeller to the air and the thrust that the air produces on the propeller [1].

Blade element theory applies the aerofoil theory to the rotating blade [1], this theory studies the forces actuating in a two-dimensional (2D) section of the blade; thus, it is necessary to have a detailed description of the geometry of the blades of the propeller. It includes both the blade chord and the pitch angle distributions along the blade span; the lift and drag coefficients are assumed to be known or can be estimated for each section of the blade. The integration of the contributions of the sections along the radius yields the total thrust and torque.

The research community has been studying quadcopters in the last years, most of them developing control systems for these vehicles, but the aerodynamic effects of the rotorcraft vehicles have been often ignored [2]. The wake of a rotorcraft has a significant effect on the overall flow field and on the performance of the vehicle. The wake is characterized by induced velocities and regions with intense vortical flow which interact with rotor blades and fuselages [3]. Using computational fluid dynamics (CFD), it is possible to analyze the rotor wake and tip vortices and also to predict the performance of the propeller.

2. Momentum theory

The simplest theory to estimate the performance of a propeller is the momentum theory. This model uses an ideal fluid, that is, the air is incompressible, not viscous and irrotational, whereas the propeller is modeled as a disc with the same area that is described by the rotation of the propeller. This disc produces a pressure jump uniformly distributed to the flow passing through it [1]. **Figure 1** shows the sketch of the model.

Applying the Bernoulli equation to the fluid before the disc,

$$p_\infty = \frac{1}{2}\rho v_i^2 + p_i \tag{1}$$

where ρ is the air density, v_i is the velocity induced by the disc to the fluid, p_i is the pressure just before the disc, and p_∞ is the pressure of the undisrupted fluid. Now applying Bernoulli after the disc,

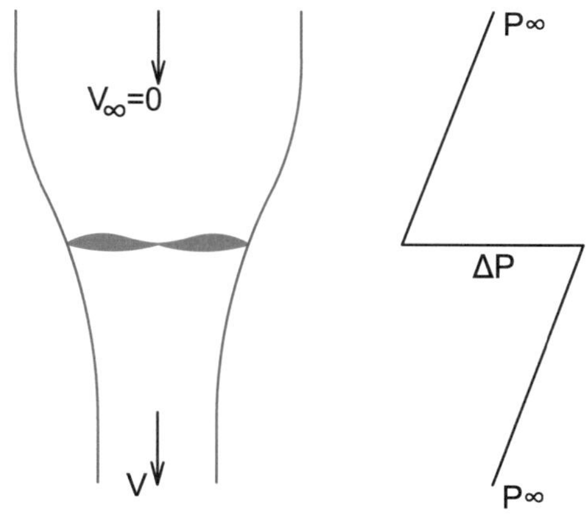

Figure 1. Disc simplification for momentum theory.

$$\frac{1}{2}\rho v_i^2 + p_i + \Delta p = p_\infty + \frac{1}{2}\rho v_\infty^2 \tag{2}$$

where v_∞ is the velocity at downstream infinity, see **Figure 1**. From Eqs. (1) and (2),

$$\Delta p = \frac{1}{2}\rho v_\infty^2 \tag{3}$$

But,

$$T = \Delta p \cdot A \tag{4}$$

where T is the thrust, Δp is the pressure jump caused by the propeller, and A is the area of the disc. From Eqs. (3) and (4),

$$T = \frac{1}{2}\rho A v_\infty^2 \tag{5}$$

Using momentum conservation, it is easy to see that the total increase on the momentum of the air must be equal to the thrust, and from **Figure 1**, $v_0 = 0$, then,

$$T = \dot{m}(v_\infty - v_0) = \rho v_i A v_\infty \tag{6}$$

Using Eq. (4),

$$\Delta p = \frac{T}{A} = \rho v_i v_\infty \tag{7}$$

And with Eq. (3),

$$\frac{1}{2}\rho v_\infty^2 = \rho v_i v_\infty \tag{8}$$

$$\frac{1}{2}v_\infty = v_i \tag{9}$$

Which means that half of the velocity induced to the fluid is done before the propeller and half after this, therefore,

$$T = 2\rho A v_i^2 \tag{10}$$

2.1. Induced power

The power induced on the air is the change in its kinetic energy per unit of time:

$$P_i = \frac{1}{2}\dot{m}\left(v_\infty^2 - v_0^2\right) = \frac{1}{2}\rho v_i A v_\infty^2 \tag{11}$$

And using Eqs. (9) and (10),

$$P_i = T v_i \tag{12}$$

$$P_i = \frac{T^{3/2}}{\sqrt{2\rho A}} \tag{13}$$

where P_i is the induced power of the rotor, that is, it is the power that the rotor induces in the air. The relation of the induced power in hover to the total power of the rotor is called *figure of merit* (FM), and it is one of the forms to express the efficiency of a rotor in hover [4]:

$$FM = \frac{P_i}{P} \tag{14}$$

Small propellers used in the UAV's are currently capable of an FM up to 0.65 [4].

3. Blade element theory

This theory analyzes aerodynamic forces on a 2D section of a blade, in order to find the contribution of each section to the total thrust; for this, it is necessary to know the chord and pitch angle distributions along the span of the blade. The *Lift Coefficient* C_L is a known data from the lift surface of each section of the blade.

Figure 2 shows a blade of a propeller with an elementary section of blade at a distance r from the axis of the propeller, with width Δr and chord c, and **Figure 3** shows the section of blade.

The angular velocity of the propeller is Ω. The velocity components in the section are as follows:

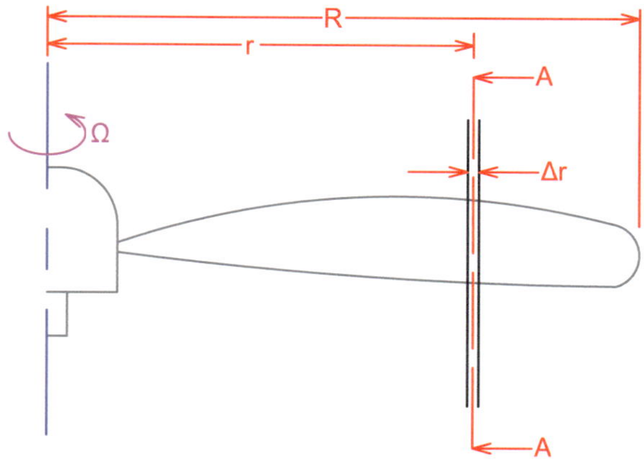

Figure 2. View of a blade.

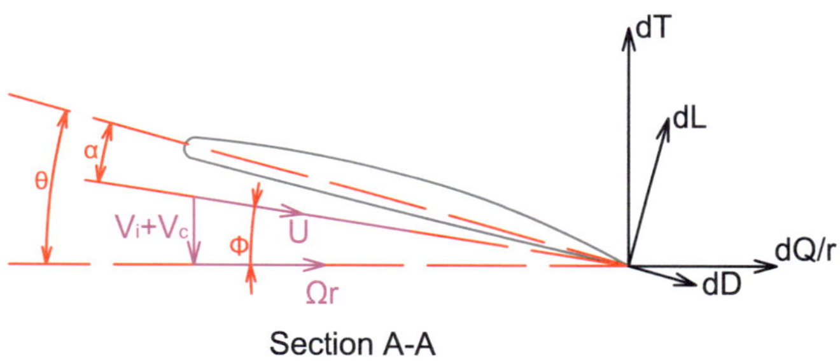

Figure 3. Details of the section of blade.

$$V_2 = \Omega r \tag{15}$$

$$V_0 = v_i + V_c \tag{16}$$

$$U = \left[(v_i + V_c)^2 + (\Omega r)^2 \right]^{1/2} \tag{17}$$

where v_i is the induced velocity and V_c is the climb velocity, see **Figure 3**. r is the radii of the section, see **Figure 2**.

The blade pitch angle is θ, and the inflow angle ϕ is calculated as

$$\phi = \tan^{-1}\left(\frac{v_i + V_c}{\Omega r}\right) \tag{18}$$

But assuming that ϕ is small, then

$$\phi = \frac{V_0}{V_2} = \frac{v_i + V_c}{\Omega r} \tag{19}$$

The incidence angle of the blade section is α

$$\alpha = \theta - \phi \tag{20}$$

For the blade section, the lift and drag forces are

$$\Delta L = \frac{1}{2}\rho U^2 c C_L \Delta r \tag{21}$$

$$\Delta D = \frac{1}{2}\rho U^2 c C_D \Delta r \tag{22}$$

where ρ is the density, c is the chord of the section, C_L and C_D are the lift and drag coefficients for the section, respectively, and Δr is the width.

From **Figure 3**,

$$\Delta T = \Delta L \cos\phi - \Delta D \sin\phi \tag{23}$$

$$\Delta Q = (\Delta L \sin\phi + \Delta D \cos\phi) r \tag{24}$$

where ΔT and ΔQ are the thrust and torque for the section of blade, respectively.

If ϕ is small,

$$\Delta T = \Delta L \cos\phi \tag{25}$$

With Eq. (21), ΔT for one blade is

$$\Delta T = \left(\frac{1}{2}\rho U^2 c C_L \Delta r\right) \cos\phi \tag{26}$$

Or for many blades

$$\Delta T = \left(\frac{1}{2}\rho U^2 c C_L \Delta r\right) B \cos\phi \tag{27}$$

where B is the number of blades of the propeller.

For a thin section, C_L could be estimated as [5]

$$C_L = 2\pi\alpha \tag{28}$$

As in the momentum theory, it is necessary to use momentum conservation in the axial direction, that is, that the change in the moment of the fluid is equal to the thrust of the blade. For a section of blade in hover,

$$\Delta T = \rho 2\pi r V_0 (V_\infty) \Delta r \tag{29}$$

But from the momentum theory, Eq. (9):

http://dx.doi.org/10.5772/intechopen.69730

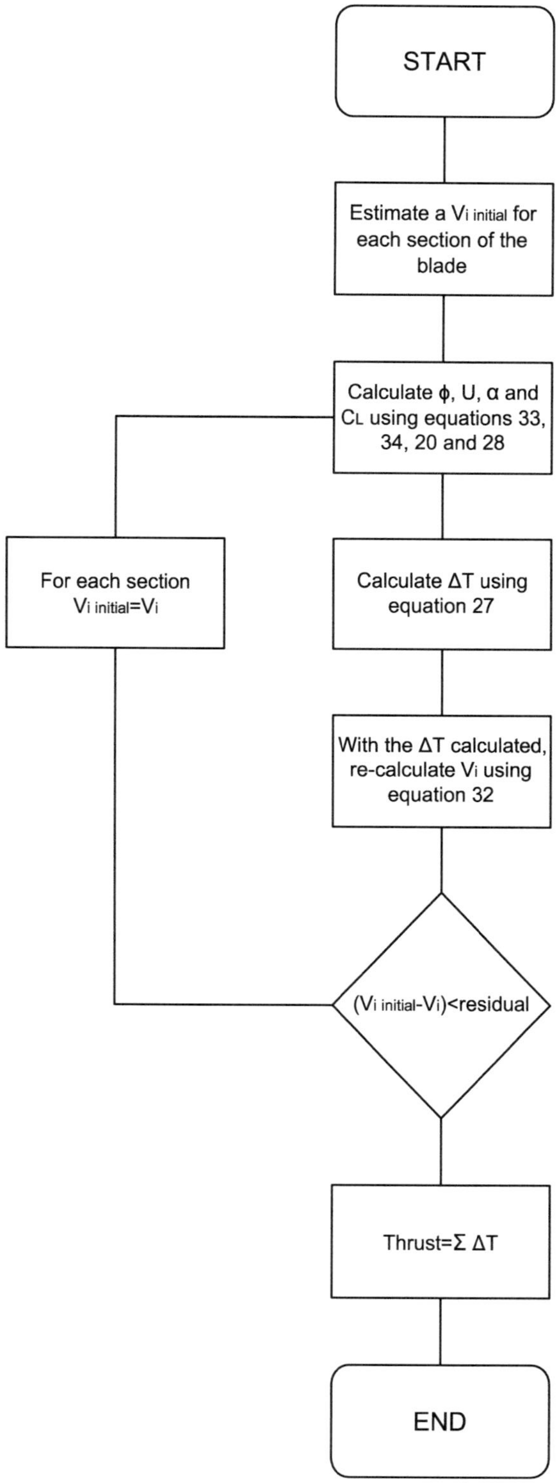

Figure 4. Method of solution using the blade element theory.

$$V_\infty = 2v_i \tag{30}$$

$$\Delta T = \rho 2\pi r V_0 (2v_i)\Delta r \tag{31}$$

But remembering that for hover $V_c = 0$, thus $V_0 = v_i$ therefore

$$\Delta T = \rho 4\pi r v_i^2 \Delta r \tag{32}$$

And Eqs. (17) and (19) become

$$U = \left[v_i^2 + (\Omega r)^2\right]^{1/2} \tag{33}$$

$$\phi = \frac{v_i}{\Omega r} \tag{34}$$

The solution method using the blade element theory requires detailed information about the geometry of the blade, that is, subdivide the blade in many sections of small width, and for each section, measure the pitch angle θ and the chord c. With Eq. (20) and Eqs. (27), (28), (32)–(34), it is possible to solve the thrust and the v_i for a propeller in hover, but it is necessary to implement an iterative method to achieve the solution. **Figure 4** shows a flow diagram describing the iterative method of solution.

4. Computational method

This study was performed using the multiple reference frame (MRF) model in steady state. The conservative equations for a MRF for an incompressible flow are [6]

$$\nabla . \vec{v}_r = 0 \tag{35}$$

$$\rho\left[\frac{\partial \vec{v}}{\partial t} + \left(\vec{v}_r \nabla \vec{v}\right)\right] + \rho\left(\vec{\Omega} \times \vec{v}\right) = -\nabla p + \nabla \bar{\bar{\tau}} + \vec{F} \tag{36}$$

where $\vec{v}_r$ is the relative velocity, $\vec{\Omega}$ is the rotational velocity (which must be constant), and in this case the Coriolis and centripetal accelerations are into a single term $\vec{\Omega} \times \vec{v}$. Many problems permit the entire computational domain to be referred to as a single rotating reference frame (SRF) [6].

Two turbulence models have been widely used to model external flows over airfoils, wings, and rotors; these models are Spalart-Allmaras and k-ω. In wall-bounded flows, the vorticity and the strain are of similar magnitude, but this is not true in a vortex core where the vorticity is high, but the strain is low. In these models, the turbulence production term is based on the velocity gradient; to include the effects of rotation, it is necessary to have some modifications to the classical models.

Many authors have studied the effect of the turbulence model used in the CFD simulations of rotors [7–10]. But though many models have been used, the Spalart-Allmaras model is the most used model to simulate rotors; this is a one equation model that solves a modeled transport equation for the turbulent viscosity [11]

$$\frac{\partial}{\partial t}(\rho\tilde{V}) + \frac{\partial}{\partial X_i}(\rho\tilde{V}U_i) = G_V + \frac{1}{\sigma\tilde{V}}\left\{\frac{\partial}{\partial X_j}\left[(\mu+\rho\tilde{V})\frac{\partial\tilde{V}}{\partial X_j}\right] + C_{b2}\left(\frac{\partial\tilde{V}}{\partial X_j}\right)^2\right\} - Y_V + S\tilde{V} \quad (37)$$

where $\tilde{v}$ is the modified turbulent viscosity, G_v is the production of turbulent viscosity, Y_v is the destruction of turbulent viscosity that occurs in the near-wall region, $\sigma\tilde{v}$ and C_{b2} are the constants of the model, v is the molecular kinematic viscosity, and $S\tilde{v}$ is a source term. More information about how to calculate the terms in Eq. (37) and the constants of the model can be found in Ref. [11].

A drawback of the Spalart-Allmaras model is that this model is insensitive to streamline curvature and system rotation [6]. A modification to the turbulence production term is available to sensitize this model to the effects of streamline curvature and system rotation. An empirical function is used as a multiplier of the production term to account for these effects; detailed information can be found in Ref. [12].

The simulations were performed using the CFD software ANSYS FLUENT v17.0. The turbulence model used in the model was the Spalart-Allmaras model with curvature correction; the SIMPLE pressure-velocity coupling solver was used, with second-order upwind discretization for convective and diffusive terms. The simulations were performed increasing the rotational velocity with small discrete steps starting with 1 rpm up to a maximum velocity of 6547 rpm. For each increase in angular velocity, the model was run until the variation of residuals was very small; in this case, the convergence criteria were adjusted to 1×10^{-6}. The working fluid is air at the conditions of the experimental test; see Section 5 of this chapter.

4.1. Geometry and computational domain

The actual rotor blade installed in the quadrotor model Araknos V2 of the Colombian company ADVECTOR is used for this study. A three-dimensional (3D) scanning was performed to obtain a CAD of the propeller; see **Figure 5**.

The diameter of the propeller (DP) is 360 mm, the root chord is 20.77 mm, the tip chord is 4.33 mm, and the chord at 75% of span is 25.31 mm. The computational domain is a vertical cylinder, with 3.3 DP in diameter and 7 DP in length; the propeller is centered in the domain and at 2.7 DP from the upper surface. The generated mesh is an unstructured grid of tetrahedral elements but with a refined grid of pyramidal elements near the propeller surface.

The simulations were performed on the server clustergate of Universidad de los Andes, with Intel X86_64 processors with a velocity of 2.4 MHz using 32 cores and 64 GB of RAM memory. The simulation for the final grid took 168 h (**Figure 6**).

Figure 5. CAD model of the propeller.

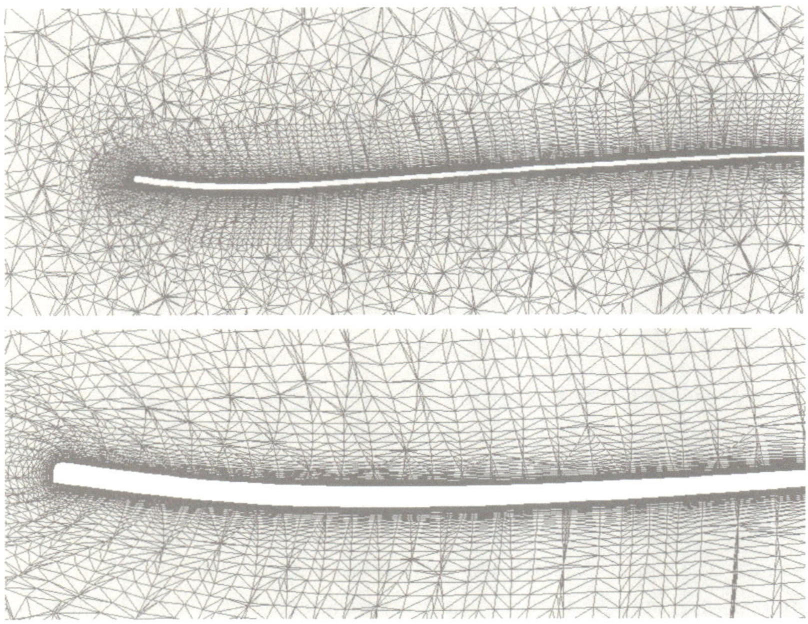

Figure 6. Mesh. Up: general mesh around the propeller; down: pyramidal elements near the wall.

A grid convergence analysis was performed using different meshes with total number of elements varying from 0.8 to 16 million cells. Total torque was used as convergence criteria at 5221 rpm. It was found that for an increase of the number of elements from 9.6 to 16.5 million, the change in the force predicted was 0.2%; it was concluded that the mesh with 9.6 million cells was sufficient.

The boundary conditions for the upper and lateral surface of the cylinder are set as pressure inlet, while the lower surface is set as a pressure outlet and the propeller boundary is a non-slip wall.

5. Experimental methodology

Flight test was performed with the quadcopter ARAKNOS V2, see **Figure 7**; the flying test was in hover at a height of 15 m from the ground, with an atmospheric pressure of 748 hPA and an average temperature of 24°C. Power for the motors was supplied only from one battery,

Figure 7. Multirotor Araknos V2 used in this study.

whereas other batteries were used as ballast. These flights were monitored using the software Mission Planner, which provided real-time information on battery voltage and control output to the engines. Flight data collected from the IMU, barometer, voltage and current sensor, and control output were recorded by the autopilot and downloaded for analysis. The data of rotational speed and electrical power recorded during hovering flight were analyzed.

The main features of the motor used in the quadcopter are as follows:

- Out runner.
- Brushless.
- kV: 530 rpm/V.
- Rotor diameter: 50 mm
- Rotor height: 10 mm.
- Shaft diameter: 3.17 mm.
- Inner resistance: 127 mΩ.
- No load current: 0.5A at 7990 rpm and 14.8 V.

6. Results

6.1. Experimental measurements

The test was performed at an average velocity of 4430 rpm, the air density was 0.87 kg/m^3, and the power measured in the motor was $P = 62.7$ W, achieving a thrust of $T = 5.89$ N.

6.2. Momentum theory

As indicated in Ref. [4], the FM for the early propellers used in UAVs was in the range of 0.4–0.45; now, the currently used propellers in these vehicles develop FM up to 0.65. With this consideration, a conservative value of $FM = 0.5$ was estimated for the propeller used in the quadcopter ARAKNOS, including the loss in the motor but without the loss of power in the coupling propeller-motor shaft, because this is a fixed coupling with screws.

From Eq. (14), it is possible to obtain the value of the induced power, taking the value of power measured in the flight test $P = 62.7$ W.

$$P_i = 31.35 \text{ W} \tag{38}$$

The diameter of the disc is 0.36 m and the air density is $\rho = 0.87 \text{ kg/m}^3$, with these data and using Eq. (13):

$$T = 5.58 \, N \tag{39}$$

The value obtained with this model is close to the measured value. It was necessary to assume the FM for the propeller.

6.3. Blade element theory

To apply the blade element theory, it was necessary to split the blade of the propeller in sections with a $\Delta r = 1$ mm, and for each section measure the chord (c) and the pitch angle (θ).

As explained in **Figure 4**, it is necessary to give an initial guess to v_i for each section of the blade; for simplicity, a uniform-induced velocity along the span of the blade was used as the initial value of v_i for the iterative process.

The initial guess of v_i is important because if the iterative method is simple, this could not converge and would be necessary to apply some techniques for convergence as, for example, Crank-Nicholson or under-relaxation [5].

A simple iterative code was written in Matlab to solve the thrust using the blade element theory, and the initial value of v_i was estimated using the momentum theory. Using the value of thrust obtained from the momentum theory and with Eq. (10),

$$v_i = 5.61 \text{ m/s} \tag{40}$$

The code needs the values of the radii, the chord, and the pitch angle for each section; also, it is necessary to indicate the rpm, in this case 4430 rpm; the density of the fluid; the width of the section; the number of blades; and the initial guess for v_i.

Adjusting the maximum residual of v_i to 1×10^{-4}, the result of the code is

$$T = 5.07 \, N \tag{41}$$

Seventy-one iterations were necessary to achieve the convergence of the method.

The thrust estimated using the blade element theory is close to the thrust measured in the experimental tests. Thought the iterative method was simple, it was easy to achieve convergence; this was possible thanks to the use of the momentum theory to estimate the initial value of v_i. From the results of the code, it was possible to see that even when the initial guess for v_i was considered as uniform along the span of the blade, the final distribution of the induced velocity is not uniform, in the root of the blade $v_i = 3.7$ m/s while at 75% of the span $v_i = 5.8$ m/s and at the tip $v_i = 2.96$ m/s.

6.4. Computational results

Figure 8 shows the wake developed by the propeller at 4500 rpm; the criterion to visualize the vortex is the Q-criterion with a threshold of 200, and the structure of the wake is similar to that described in Ref. [1], where it is clear that the strong vortex is created at the tip of the blade, also the tip vortex and the inner vortex descend below the rotor following a helical path.

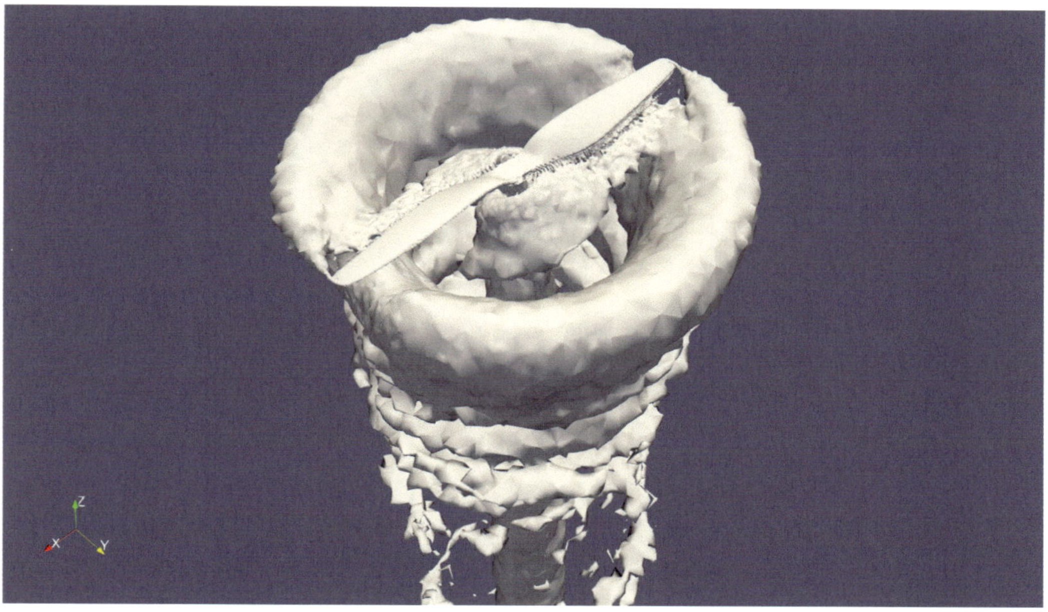

Figure 8. Vortex Contours at 4500 rpm using Q-criterion with a threshold of 200.

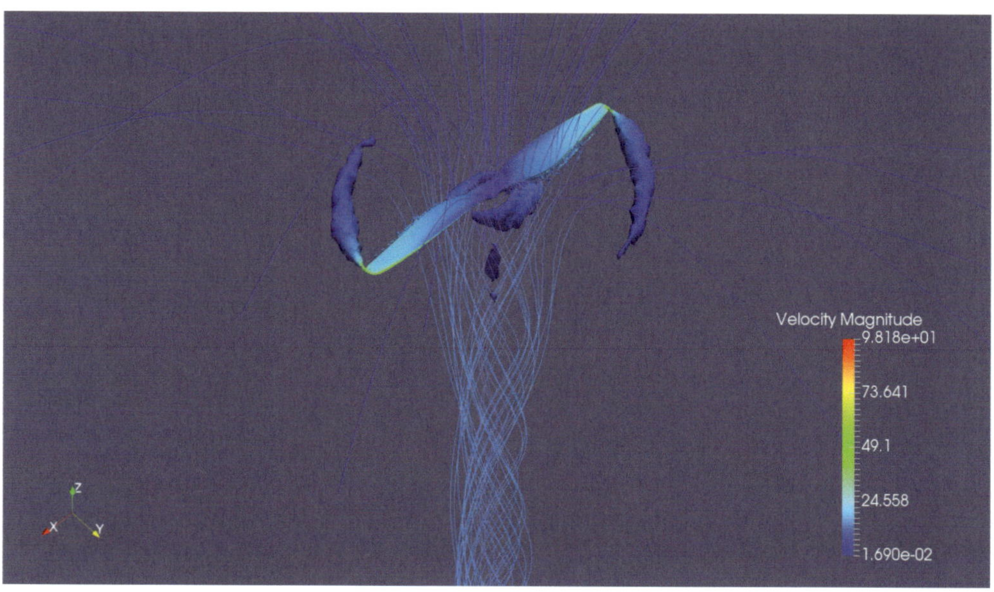

Figure 9. Streamlines colored by velocity magnitude and contours of Q-criterion = 50,000, at 4500 rpm.

There are various methods used to visualize vortex; some of these methods are based on the tensor gradient of velocity as, for example, the *Q*-criterion and λ_2 -criterion; other methods are based on vorticity, for example, the magnitude of vorticity, there are some Lagrangian methods as direct Lyapunov exponents (DLE) or the Mz criterion; more information about the methods to visualize vortex structures could be found in Ref. [13].

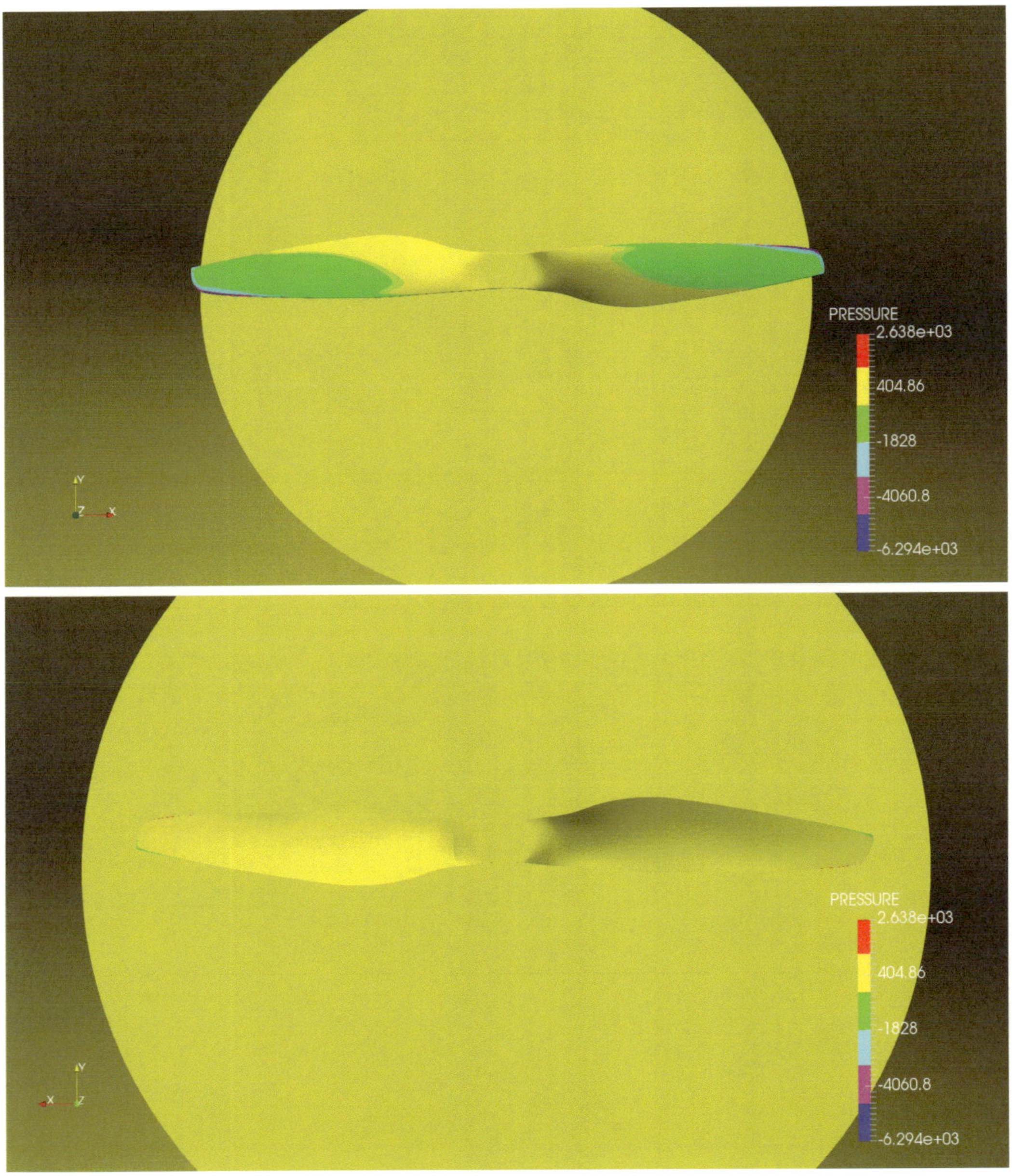

Figure 10. Contours of pressure on the surfaces of the propeller in PA at 4500 rpm, up upper surface, down lower surface.

Figure 9 shows streamlines colored by velocity magnitude and the vortex contours with a threshold for Q-criterion of 50,000; the model shows the increase of the magnitude of the velocity down the propeller, also the streamlines swirl around the axis of the propeller.

Figure 10 shows the contours of pressure for the upper and lower surfaces of the propeller. As expected, the pressure is lower at the upper surface of the propeller.

Finally, the thrust predicted by the model is

$$T = 6.99\ N \tag{42}$$

The model is able to predict the wake, pressure, velocity fields, and thrust of the propeller at hover and for different operational rotational velocities using the MRF method in steady state. The model overpredicts the thrust; this difference could be due to the interaction of the four rotors between them and also with the fuselage of the aircraft in the experimental measurements.

7. Conclusions

The momentum and blade element theories are useful tools to achieve a preliminary estimation on the performance of a propeller in hover, though it is important to understand the strong simplifications that each of these theories imply. For the momentum theory, simplifications are strong, for example, the flow is ideal, the simplification of the disc implies that the propeller has an infinite number of blades, also the induced velocity is uniform along the span of the propeller, and for the solution it was necessary to estimate the efficiency of the propeller in hover; this could lead to unrealistic results. For the blade element theory, the first simplification is that the flow is studied as a 2D flow, and effects of a 3D flow are ignored, for example, the induced velocities by the tip vortex, also linear aerofoil properties are used for the sections of the blade, that is, to estimate C_L. The method does not simplify the induced velocity as being uniform along the blade; instead it is possible to estimate the induced velocity distribution for the propeller in hover.

A CFD model was implemented to study the flow around a propeller in hover. The model is able to predict the wake, pressure, velocity fields, and thrust using the MRF method in steady state. The predicted wake is in agreement with the literature.

The theoretical approximations for thrust are closer to the measurement of the experimental test, while the computational method overestimates the thrust. **Table 1** shows that the maximum percentage of error is 18.7% for the CFD simulation.

For a better validation of the models analyzed in this chapter, a test bench to a more accurate measurement of thrust is necessary.

Method	Experimental	Momentum theory	Blade theory	Computational simulation
Thrust [N]	5.89	5.58	5.07	6.99
Error		5.3%	13.9%	18.7%

Table 1. Comparison of the results with the experimental measurement.

The control of a quadcopter is based on the thrust and torque of the propulsion system; therefore, an adequate estimation of the aerodynamic parameters is very important for the control of the vehicle. The methods described have some characteristics that are important to apply in some of the stages of the design, for example, the moment theory is important for a first estimation of the size of the rotor; **Table 2** shows the advantages and disadvantages of each method.

Method	Advantages	Disadvantages
Momentum theory	• Simple method • First estimation of propeller aerodynamic performance (thrust torque) • With this approximation it is possible to size a rotor for a given power	• Strong suppositions • It does not account for number of blades • It does not concern the details of the flow around the blades • It is not useful to design the rotor because it does not account the shape of the blade • It is an ideal fluid theory
Blade theory	• The model allows studying the shape of the blade, and therefore it possible to design the rotor • The mathematical solution is no complex and not computationally expensive	• The model simplifies the blade to 2D non-interacting sections • It is necessary to estimate the C_L distribution of the blade • It is an ideal fluid theory
Computational simulation	• It is a real fluid method • It is possible for a more detailed study of the structure of the flow • It is a 3D method • It allows the verification of the results of the other methods	• High computational cost • It is a complex method

Table 2. Benefits and limitations of each method.

Acknowledgements

The authors like to thank the Departamento Administrativo de Ciencia, Tecnología e Innovación Departamento (COLCIENCIAS) and its program to sponsor PhD students in Colombia, also to the high-performance computing center at Universidad de los Andes for providing the computational resources, and ADVECTOR for being part of this research, sharing their technical data and to perform the necessary flying tests.

Nomenclature

A	Area of the disc
c	Chord
C_D	Drag coefficient
C_L	Lift coefficient
D	Drag
FM	Figure of merit
L	Lift
P_i	Induced power of the rotor

http://dx.doi.org/10.5772/intechopen.69730

p_i	Pressure just before the disc
p_∞	Pressure of the undisrupted fluid
Q	Torque
T	Thrust
V_c	Climb velocity
v_i	Induced velocity
v_∞	Velocity at downsteam infinity
Δp	Pressure jump caused by the propeller
ρ	Density
θ	Blade pitch angle
ϕ	Inflow angle
α	Incidence angle

Appendix

Matlab code for blade element theory.

```
%**********************************************************
%% It is necessary to load the vectors of
% Radii r[ m] , Chord c[ m] and Pitch Angle theta[ rad]
%**********************************************************
%**********************************************************
%% Data
%**********************************************************
rpm=4430;% Rotational velocity rpm
omega=rpm*2*pi/60;% Rotational velocity[ rad/s]
rho=0.87;% Density[ kg/m^3]
dr=0.001;% Delta r[ m]
B=2;% # blades
Vi=5.61;% Initial guess for Vi[ m/s]
%**********************************************************
%% Initial Velocities
%**********************************************************
vi0=zeros(length (r),1);
vi0=vi0+Vi;% Vector of initial vi[ m/s]
v2=r.*omega;% Vector of V2[ m/s]
%**********************************************************
%% Iterative method
%**********************************************************
Rmax=10;% Residual
n=0;% Iteration Counter
while Rmax>0.0001;
```

```
U=((vi0.^2)+(v2.^2)).^0.5;% U[m/s]
fi=vi0./v2;% Angle fi[rad]
alfa=theta-fi;% Angle of inflow[rad]
Cl=alfa.*(2*pi);% Lift Coefficient
DT=((((U.^2)*0.5*rho).*c).*Cl).*dr*B).*cos(fi);% Differential of Thrust
[N]
vi=((DT./(rho*4*pi*dr))./r).^0.5;% New Vi[m/s]
R=vi0-vi;% Residual of Vi
R=abs(R);
Rmax=max(R);
n=n+1;
vi0=vi;
end
THRUST=sum(DT)
Residual=Rmax
Iteraciones=n
```

Author details

Andres Mauricio Pérez Gordillo

Address all correspondence to: am.perez259@uniandes.edu.co

Universidad de los Andes, Bogotá, Colombia

References

[1] Seddon J. Rotor in vertical flight: Momentum theory and wake analysis. In: Basic Helicopter Aerodynamics. 1st ed. BSP Professional Books, Great Britain: 1990

[2] Hoffmann G, Huang H, Waslander S, Tomlin C. Quadrotor helicopter flight dynamics and control: Theory and experiment. AIAA Guidance, Navigation and Control Conferece and Exhibit. 20 - 23 August 2007, Hilton Head, South Carolina. AIAA 2007–6461

[3] Potsdam M, Pulliam T. Turbulence Modeling Treatment for Rotorcraft Wakes. US Army Research, Development, and Engineering Command, NASA Ames Research Center. AHS specialists conference on aeromechanics. San Francisco, CA, 2008

[4] Pereira JL. Hover and wind-tunnel testing of shrouded rotors for improved micro air vehicle design [dissertation]. University of Maryland, College Park; 2008. p. 349

[5] Auld, Srinivas. Aerodynamics for Students [Internet]. 1995–2016. Available from: http://s6.aeromech.usyd.edu.au/aerodynamics/index.php/sample-page/propulsion/blade-element-rotor-theory/ [Accessed: January 10, 2017]

[6] ANSYS Inc. ANSYS FLUENT 12.0/12.1 Documentation [Internet] 2009. Available from: http://www.afs.enea.it/project/neptunius/docs/fluent/index.htm [Accessed: September 20, 2016]

[7] Lazaro C, Poroseva S. Computational analysis of the blade number effect on the performance of a ducted propeller. In: AIAA Atmospheric Flight Mechanics Conference; January 5–9, 2015; Kissimmee, FL: American Institute of Aeronautics and Astronautics; 2015

[8] Gomez S, Gilkey L, Kaiser B, Poroseva S. Computational analysis of a tip vortex structure shed from a bio-inspired blade. In: AIAA Applied Aerodynamics Conference; June 16–20, 2014; Atlanta: American Institute of Aeronautics and Astronautics; 2014

[9] Duraisamy K, Baeder JD. High resolution wake capturing methodology for hovering rotors. Journal of the American Helicopter Society. 2007;52(2):110–122

[10] Doerff P, Szulc O. Numerical Simulation of Model Helicopter Rotor in Hover. Institute of Fluid-Flow Machinery PAS. TASK Quarterly : scientific bulletin of Academic Computer Centre in Gdansk. 2008;12(3-4): 227–236

[11] Spalart P, Allmaras S. A One-Equation Turbulence Model for Aerodynamic Flows. American Institute of Aeronautics and Astronautics Technical Report; 30th Aerospace Sciences Meeting and Exhibit, Reno, NV, U.S.A., 1992

[12] Spalart P, Shur ML. On the sensitization of turbulence models to rotation and curvature. Aerospace Science and Technology. 1997;1(5): 297–302

[13] Holmen V. Methods for vortex identification [dissertation]. Master's Theses in Mathematical Sciences. Mathematics (Faculty of Technology) and Numerical Analysis. lund university libraries 2012. p. 46

Section 2

Dynamics and Control

Chapter 5

Nonlinear Dynamics and Control of Aerial Robots

Mahmut Reyhanoglu and Muhammad Rehan

Additional information is available at the end of the chapter

http://dx.doi.org/10.5772/intechopen.69641

Abstract

Aerial robotics is one of the fastest growing industry and has a number of evolving applications. Higher agility make aerial robots ideal candidate for applications like rescue missions especially in difficult to access areas. This chapter first derives the complete nonlinear dynamics of an aerial robot consisting of a quadcopter with a two-link robot manipulator. Precise control of such an aerial robot is a challenging task due to the fact that the translational and rotational dynamics of the quadcopter are strongly coupled with the dynamics of the manipulator. We extend our previous results on the control of quadrotor UAVs to the control of aerial robots. In particular, we design a backstepping and Lyapunov-based nonlinear feedback control law that achieves point-to-point control of the areal robot. The effectiveness of this feedback control law is illustrated through a simulation example.

Keywords: quadcopter, robot manipulator, backstepping method, nonlinear control

1. Introduction

The recent surge of interest in applications involving unmanned aerial vehicles (UAVs) has inspired several research efforts in UAV dynamic modeling and control. In particular, nonlinear control of fixed-wing UAVs has attracted considerable research efforts during recent years both for civilian and military purposes. The control approaches developed for fixed-wing UAVs include gain scheduling, model predictive control, backstepping, sliding mode, nested saturation, fuzzy control, H_∞ control, dynamic inversion based control, model reference adaptive control, and model based fault tolerant control [1–12].

While control applications involving fixed-wing UAVs have been widely investigated in recent literature, quadrotor UAV (quadcopter) control applications are growing in popularity due to their maneuverability and versatility. Quadcopters offer practical advantages over fixed-wing

INTECH
open science | open minds

© 2017 The Author(s). Licensee InTech. This chapter is distributed under the terms of the Creative Commons Attribution License (http://creativecommons.org/licenses/by/3.0), which permits unrestricted use, distribution, and reproduction in any medium, provided the original work is properly cited.

UAVs in military and civilian applications involving search and rescue, area mapping, and surveillance. However, actuator constraints, sensor limitations, and the high degree of nonlinearity and uncertainty inherent in the system dynamics present specific challenges in control system design for quadcopters. Linear control approaches, such as PID, LQR or LQG are popularly utilized to address the quadcopter control problem [13–15]. Although linear control methods have been shown to perform well in their respective quadcopter control tasks, the effectiveness of linear control methods can only be guaranteed over a limited range of operating conditions. The highly agile nature of quadcopters necessitates the development of control methods that can be applied over a wide range of time-varying, uncertain, and potentially adversarial operating conditions. To achieve reliable quadcopter control over a wider operational envelope, several nonlinear control methods have recently been presented. Popular nonlinear control methods for quadrotor systems include backstepping, feedback linearization, dynamic inversion, adaptive control, Lyapunov-based robust control, fuzzy-model approach, and sliding mode control [16–19]. In Ref. [16], a passivity-based quaternion feedback control strategy is presented for a hover system (quadrotor UAV test bed), which achieves asymptotic attitude regulation. The proposed control design incorporates the input voltage constraints inherent in practical UAV systems. A rigorous Lyapunov-based analysis is provided to prove asymptotic regulation of the hover system attitude to a desired set point. In Ref. [17], a sliding mode control (SMC) strategy is presented for a quadrotor-based hover system, which achieves asymptotic attitude regulation in the presence of electrical and physical constraints. A sliding mode observer is employed to estimate the angular velocities. In addition, the proposed control design incorporates the input voltage constraints inherent in practical systems. A rigorous Lyapunov-based analysis is provided to prove asymptotic regulation of the hover system attitude to a desired set point.

Aerial robotics is one of the fastest growing industry and has a number of evolving applications. Higher agility make aerial robots ideal candidate for applications like rescue missions especially in difficult to access areas. Furthermore, swarm robotics (multiple robot working together) is another exciting application of the aerial robotics, for example coordinated assembly at higher altitudes. These robots can behave like individuals working in a group without centralized control. Researchers have developed intelligent control algorithms for the swarms after deep study of animal behavior in herds, bird flocks and fish schools. In some applications, for an aerial robot, linear control theory works well but these control techniques are effective in a limited operating regions. Moreover, the motion of arm induces disturbances to the quadcopter dynamics so the linear controllers lose their effectiveness during operation and sometime the closed loop system becomes unstable. In order to accomplish complex missions in presence of uncertainties in the environment, to achieve better maneuverability and precise 3D position and attitude control, nonlinear control techniques have been found effective [20–27], In Ref. [20] a set of nonlinear control laws have been proposed for aerial manipulator that provide asymptotic attitude and position tracking. Backstepping-based nonlinear control scheme for automatic trajectory tracking for aerial manipulators has been proposed in Refs. [22, 24].

In this chapter, we extend our results on the control of quadcopters to the control of aerial robots. We derive the complete nonlinear dynamics of an aerial robot consisting of a quadcopter with a two-link robot manipulator. Precise control of such a robot is a challenging task because attitude and position dynamics of the quadcopter are strongly coupled with the dynamics of the manipulator. We develop nonlinear control laws that ensure the control of position and attitude of the aerial robot. Simulation results are included to demonstrate the effectiveness of the control laws.

2. Mathematical model

This section formulates the dynamics of an aerial robot consisting of a quadcopter with a two-link robot manipulator. The quadcopter is represented as a base body and the links as internal bodies. The equations of motion are expressed in terms of the three dimensional translational velocity vector, the attitude, the angular velocity, and the internal (shape) coordinates representing the configuration of the two links.

2.1. Multibody vehicle dynamics

Following the development in [28], let $v \in \mathbb{R}^3$, $\omega \in \mathbb{R}^3$, and $\eta \in \mathbb{S}^1 \times \mathbb{S}^1$ denote the base body translational velocity vector, the base body angular velocity vector, and the vector of internal coordinates, respectively. In these variables, the kinetic energy has the form $T = T(v, \omega, \eta, \dot{\eta})$, which is $SE(3)$-invariant in the sense that it does not depend on the base body position and attitude. The equations of motion of the quadrotor with internal dynamics are shown to be given by:

$$\frac{d}{dt}\frac{\partial T}{\partial v} + \hat{\omega}\frac{\partial T}{\partial v} = F_t, \tag{1}$$

$$\frac{d}{dt}\frac{\partial T}{\partial \omega} + \hat{\omega}\frac{\partial T}{\partial \omega} + \hat{v}\frac{\partial T}{\partial v} = \tau_r, \tag{2}$$

$$\frac{d}{dt}\frac{\partial T}{\partial \dot{\eta}} - \frac{\partial T}{\partial \eta} = \tau_s, \tag{3}$$

where $F_t \in \mathbb{R}^3$, $\tau_r \in \mathbb{R}^3$ denote the vectors of generalized control forces and generalized control torques, respectively, that act on the base body, and $\tau_s \in \mathbb{R}^3$ is the vector of joint torques. For a given vector $a = [a_1\ a_2\ a_3]^T \in \mathbb{R}^3$, the skew-symmetric matrix $\hat{a}$ defines the corresponding cross-product operation $a\times$, given by

$$\hat{a} = \begin{bmatrix} 0 & -a_3 & a_2 \\ a_3 & 0 & -a_1 \\ -a_2 & a_1 & 0 \end{bmatrix}. \tag{4}$$

Note that Eqs. (1) and (2) are identical to Kirchhoff's equations [29], which can also be expressed in the form of Euler-Poincaré equations.

2.2. Nonlinear equations of motion

Consider an aerial robot that consists of a quadcopter with a two DOF manipulator arm moving in a three-dimensional space as shown in **Figure 1**, where $p = [x\ y\ z]^T$ denotes the

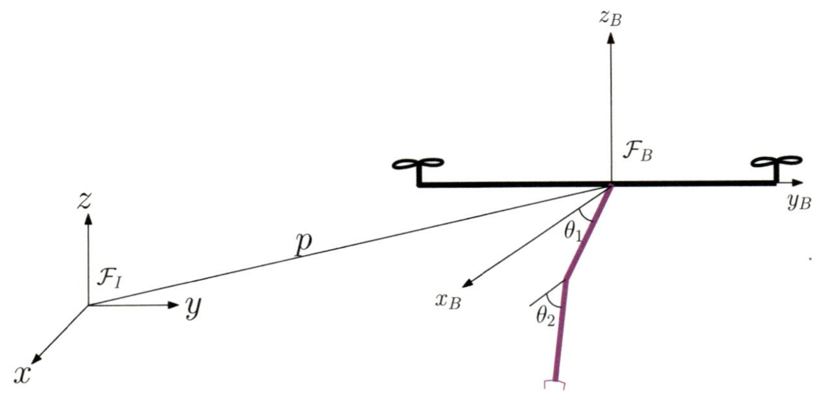

Figure 1. Model of a quadcopter with a robotic arm.

inertial position of the center of mass of the quadcopter. Denote by xyz axes the inertial frame $\mathcal{F}_1$ and by $x_B\, y_B\, z_B$ axes the body-fixed frame $\mathcal{F}_B$ with the origin at the CM of the quadcopter. Let R denote the attitude matrix of the quadcopter and (v, ω) be the translational and angular velocities of the CM of the quadrotor in $\mathcal{F}_B$. Then, the translational and rotational kinematics can be expressed as

$$\dot{p} = Rv, \tag{5}$$

$$\dot{R} = R\hat{\omega}. \tag{6}$$

The quadcopter consists of four propellers connected to a rigid frame. Each propeller is mounted on the frame at a distance l from the origin. The quadcopter has a mass m and inertia matrix J defined with respect to the axes of rotation. Due to symmetry of the system, J is diagonal, that is, $J = \text{diag}\{J_{xx}, J_{yy}, J_{zz}\}$. We refer to rotation about the x_B-axis as roll, rotation about the y_B-axis as pitch, and rotation about the z_B-axis as yaw. The propellers generate lift forces

$$F_i = b\Omega_i^2 = bK_v^2 V_i^2, \tag{7}$$

where Ω_i, V_i denote, respectively, the angular rate and input voltage for propeller i, and b is the thrust coefficient. The total thrust is given by

$$F_p = \sum_{i=1}^{4} F_i e_3 = bK_v^2 (V_1^2 + V_2^2 + V_3^2 + V_4^2) e_3, \tag{8}$$

where $e_3 = [0\ 0\ 1]^T \in \mathbb{R}^3$ is the third standard basis vector.

Note that, as shown in **Figure 2**, propellers 1 and 3 rotate clockwise, and propellers 2 and 4 rotate counter-clockwise. By balancing the torque between opposing propellers, the roll and pitch angle can be controlled. Since all four propellers generate a net torque about the yaw axis, the yaw angle can be controlled by balancing the torque generated by clockwise and counter-clockwise rotating propellers.

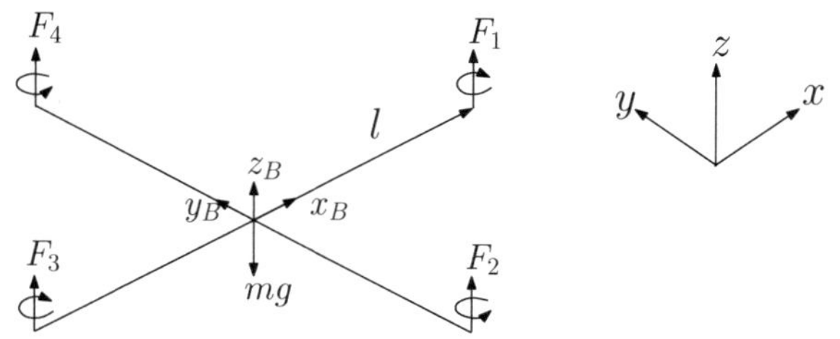

Figure 2. Model of the quadcopter.

The arm is attached at the CM of the quadcopter and it can only move in xz-plane of the body-fixed frame F_B. The physical constants are the quadcopter mass m, the link masses m_i, $i = 1,2$, and the payload mass m_p. Let l_{c_i} denote the distance from joint i to the CM of link i and l_i be the length of link i. The position vectors for the CM of the links and the payload with respect to the CM of the base body in F_B can be written as

$$\rho_1 = [l_{c_1}\cos\theta_1 \quad 0 \quad l_{c_1}\sin\theta_1]^T, \tag{9}$$

$$\rho_2 = [l_1\cos\theta_1 + l_{c_2}\cos\theta_2 \quad 0 \quad l_1\sin\theta_1 + l_{c_2}\sin\theta_2]^T, \tag{10}$$

$$\rho_p = [l_1\cos\theta_1 + l_2\cos\theta_2 \quad 0 \quad l_1\sin\theta_1 + l_2\sin\theta_2]^T. \tag{11}$$

Let $F_t = F_g + F_p$, where F_g and F_p denote, respectively, the gravitational force acting on aerial robot and the total thrust generated by the four propellers. Also let $\tau_r = \tau_g + \tau_p$, where τ_g and τ_p are the torque acting on the aerial robot due to gravity and the torque generated by the propellers, respectively.

Clearly, F_g and τ_g can be computed as

$$F_g = -m_t g R^T e_3, \tag{12}$$

$$\tau_g = -g[m_1\rho_1 + m_2\rho_2 + m_p\rho_p] \times R^T e_3, \tag{13}$$

where $m_t = m + m_1 + m_2 + m_p$.

The generalized torque vector τ_p (expressed in the body frame) comprises the following components:

- Propellers 2 and 4 generate a moment $l(F_4 - F_2) = blK_v^2(V_4^2 - V_2^2)$ about the roll axis.
- Propellers 1 and 3 generate a moment $l(F_3 - F_1) = blK_v^2(V_3^2 - V_1^2)$ about the pitch axis.
- The sum of all torques about z-axis is $dK_v^2(V_1^2 - V_2^2 + V_3^2 - V_4^2)$ and causes a yaw moment.
- The rotation of the propellers causes the gyroscopic effect $J_r K_v \omega_y(V_1 - V_2 + V_3 - V_4)$ about the roll-axis and $-J_r K_v \omega_x(V_1 - V_2 + V_3 - V_4)$ about the pitch-axis.

Here d denotes the drag coefficient, l is the distance from the pivot to the motor, and J_r is the rotor inertia. Combined, the generalized torque τ_p can be expressed as

$$\tau_p = \begin{bmatrix} blK_v^2(V_4^2 - V_2^2) + J_r\omega_y\Omega_r \\ blK_v^2(V_3^2 - V_1^2) - J_r\omega_x\Omega_r \\ dK_v^2(V_1^2 - V_2^2 + V_3^2 - V_4^2) \end{bmatrix}, \tag{14}$$

where $\Omega_r := K_v(V_1 - V_2 + V_3 - V_4)$ is the overall residual angular speed of the propellers.

Let $\eta = [\theta_1\ \theta_2]^T$ denote the shape variables. Then, the linear and angular velocities of each link, and the linear velocity of the payload can be expressed in $\mathcal{F}_B$ as

$$v_i = v + \hat{\omega}\rho_i + \frac{\partial \rho_i}{\partial \eta}\dot{\eta} = v - \hat{\rho}_i\omega + \frac{\partial \rho_i}{\partial \eta}\dot{\eta},\ i = 1, 2, \tag{15}$$

$$\omega_i = \omega + \dot{\theta}_i e_2 = \omega + C_i(\eta)\dot{\eta},\ i = 1, 2, \tag{16}$$

$$v_p = v + \hat{\omega}\rho_p + \frac{\partial \rho_p}{\partial \eta}\dot{\eta} = v - \hat{\rho}_p\omega + \frac{\partial \rho_p}{\partial \eta}\dot{\eta}, \tag{17}$$

where $e_2 = [0\ 1\ 0]^T \in \mathbb{R}^3$ is the second standard basis vector. The total kinetic energy can now be expressed as

$$T(v, \omega, \eta, \dot{\eta}) = \frac{1}{2}mv^T v + \frac{1}{2}\omega^T J\omega + \frac{1}{2}\sum_{i=1}^{2}(m_i {v_i}^T v_i + {\omega_i}^T J_i \omega_i) + \frac{1}{2}m_p {v_p}^2, \tag{18}$$

where $J_i = R_i^T \bar{J}_i R_i$ is the inertia matrix of the ith link with respect the body frame F_B, R_i is the rotation matrix of the ith link, which is given by

$$R_i = \begin{bmatrix} \cos\theta_i & 0 & \sin\theta_i \\ 0 & 1 & 0 \\ -\sin\theta_i & 0 & \cos\theta_i \end{bmatrix},\quad i = 1, 2, \tag{19}$$

and $\bar{J}_i$ denotes the inertia matrix of the ith link with respect to $x_i y_i z_i$-axes attached to the link. Assuming the two links are made up of homogeneous rods, $\bar{J}_i$ can be expressed as

$$\bar{J}_i = \begin{bmatrix} 0 & 0 & 0 \\ 0 & \frac{1}{12}m_i l_i^2 & 0 \\ 0 & 0 & \frac{1}{12}m_i l_i^2 \end{bmatrix}. \tag{20}$$

Applying Kirchhoff's equations (1) and (2), the complete nonlinear equations of motion can be obtained as

$$\begin{bmatrix} M & K & B_t \\ K^T & \bar{J} & B_r \\ B_t^T & B_r^T & \overline{m} \end{bmatrix} \begin{bmatrix} \dot{v} \\ \dot{w} \\ \ddot{\eta} \end{bmatrix} = \begin{bmatrix} F_t \\ \tau_r \\ \tau_s \end{bmatrix} - \begin{bmatrix} \hat{w}M & \dot{K} + \hat{w}K & \dot{B}_t + \hat{w}B_t \\ \dot{K}^T + \hat{w}K^T + \hat{v}M & \dot{\bar{J}} + \hat{w}\bar{J} + \hat{v}K & \dot{B}_a + \hat{w}B_a + \hat{v}B_t \\ \dot{B}_t^T & \dot{B}_a^T & \dot{\overline{m}} \end{bmatrix} \begin{bmatrix} v \\ w \\ \dot{\eta} \end{bmatrix} + \begin{bmatrix} 0 \\ 0 \\ \frac{\partial L}{\partial \eta} \end{bmatrix}, \tag{21}$$

where

$$M = m_t I_{3\times 3}, \tag{22}$$

$$\bar{J} = J + m_1 \hat{\rho}_1^T \hat{\rho}_1 + m_2 \hat{\rho}_2^T \hat{\rho}_2 + m_p \hat{\rho}_p^T \hat{\rho}_p + J_1 + J_2, \tag{23}$$

$$\overline{m} = m_1 \frac{\partial \rho_1}{\partial \eta}^T \frac{\partial \rho_1}{\partial \eta} + m_2 \frac{\partial \rho_2}{\partial \eta}^T \frac{\partial \rho_2}{\partial \eta} + m_p \frac{\partial \rho_p}{\partial \eta}^T \frac{\partial \rho_p}{\partial \eta} + C_1^T J_1 C_1 + C_2^T J_2 C_2, \tag{24}$$

$$K = -m_1 \hat{\rho} 1 - m_2 \hat{\rho}_2 - m_p \hat{\rho}_p, \tag{25}$$

$$B_t = m_1 \frac{\partial \rho_1}{\partial \eta} + m_2 \frac{\partial \rho_2}{\partial \eta} + m_p \frac{\partial \rho_p}{\partial \eta}, \tag{26}$$

$$B_r = m_1 \hat{\rho}_1 \frac{\partial \rho_1}{\partial \eta} + m_2 \hat{\rho}_2 \frac{\partial \rho_2}{\partial \eta} + m_p \hat{\rho}_p \frac{\partial \rho_p}{\partial \eta} + J_1 C_1 + J_2 C_2. \tag{27}$$

Complete description of the above coefficient matrices are given in the appendix. The objective is to simultaneously control the 6 DOF motion of the quadcopter and the 2 DOF internal dynamics of the robot arm using only 4 propellers and 2 joint torque motors. In this regard, equations of motion given by (21) represents an interesting example of underactuated mechanical systems. In our previous research [30–32], we have developed theoretical controllability and stabilizability results for a large class of underactuated mechanical systems using tools from nonlinear control theory. We have also developed effective nonlinear control design methodologies [32] that we applied to several examples of underactuated mechanical systems, including underactuated space vehicles [33].

3. Nonlinear control design

The translational and rotational dynamics of the quadcopter are coupled with the dynamics of its robotic arm; this makes controller design very complicated. The equations of motion in component form are given by

$$M\dot{v} + K\dot{\omega} + B_t\ddot{\eta} = F_t - \hat{\omega}Mv - (\dot{K} + \hat{\omega}K)\omega - (\dot{B}_t + \hat{\omega}B_t)\dot{\eta}, \tag{28}$$

$$K^T\dot{v} + \bar{J}\dot{\omega} + B_a\ddot{\eta} = \tau_r - (\dot{K}^T + \hat{\omega}K^T + \hat{v}M)v - (\dot{\bar{J}} + \hat{\omega}\bar{J} + \hat{v}K)\omega - (\dot{B}_a + \hat{\omega}B_a + \hat{v}B_t)\dot{\eta}, \tag{29}$$

$$B_t^T \dot{\nu} + B_a^T \dot{\omega} + \overline{m}\ddot{\eta} = \tau_s - \dot{B}_t^T \nu - \dot{B}_a^T \omega - \dot{\overline{m}}\dot{\eta} + \frac{\partial L}{\partial \eta}. \tag{30}$$

Eq. (28) can be rewritten as

$$M\dot{\nu} = F_t - \hat{\omega} M \nu + F_d, \tag{31}$$

where

$$F_d = -K\dot{\omega} - B_t \ddot{\eta} - (\dot{K} + \hat{\omega} K)\omega - (\dot{B}_t + \hat{w} B_t)\dot{\eta}. \tag{32}$$

Eq. (31) can be simplified as

$$\begin{bmatrix} \ddot{x} \\ \ddot{y} \\ \ddot{z} \end{bmatrix} = \begin{bmatrix} 0 \\ 0 \\ -g \end{bmatrix} + R\hat{e}_3 u_1 + \overline{F}_d. \tag{33}$$

where $\overline{F}_d = F_d/m_t$ and

$$u_1 = bK_v^2(V_1^2 + V_2^2 + V_3^2 + V_4^2)/m_t, \tag{34}$$

Similarly, Eq. (29) can be rewritten as

$$\overline{J}\dot{\omega} = \tau_r - (\dot{\overline{J}} + \hat{\omega}\overline{J})\omega + \tau_d, \tag{35}$$

where

$$\tau_d = -K^T \dot{v} - B_a \ddot{\eta} - (\dot{K}^T + \hat{\omega} K^T + \hat{v} M)\nu - \hat{v} K \omega - (\dot{B}_a + \hat{\omega} B_a + \hat{v} B_t)\dot{\eta}. \tag{36}$$

Eq. (35) can be simplified as

$$\dot{\omega} = \overline{J}^{-1}\tau_r - \overline{J}^{-1}(\dot{\overline{J}} + \hat{\omega}\overline{J})\omega + \overline{\tau}_d, \tag{37}$$

where $\overline{\tau}_d = \overline{J}^{-1}\tau_d$.

Ignoring $\overline{F}_d$ and $\overline{\tau}_d$ in Eqs. (33) and (37), equations of motion can be expressed as

$$\ddot{x} = (\cos\phi \sin\theta \cos\Psi + \sin\phi \sin\Psi)u_1, \tag{38}$$

$$\ddot{y} = (\cos\phi \sin\theta \sin\Psi - \sin\phi \cos\Psi)u_1, \tag{39}$$

$$\ddot{z} = -g + (\cos\phi \cos\theta)u_1, \tag{40}$$

$$\begin{bmatrix} \dot{\omega}_x \\ \dot{\omega}_y \\ \dot{\omega}_z \end{bmatrix} = \overline{J}^{-1}\left(\tau_g + \begin{bmatrix} J_r \omega_y \Omega_r \\ -J_r \omega_x \Omega_r \\ 0 \end{bmatrix} - (\dot{\overline{J}} + \hat{\omega}\overline{J})\omega\right) + \begin{bmatrix} u_2 \\ u_3 \\ u_4 \end{bmatrix}, \tag{41}$$

where

$$\begin{bmatrix} u_2 \\ u_3 \\ u_4 \end{bmatrix} = \bar{J}^{-1} \begin{bmatrix} blK_v^2(V_4^2 - V_2^2) \\ blK_v^2(V_3^2 - V_1^2) \\ blK_v^2(V_1^2 - V_2^2 + V_3^2 - V_4^2) \end{bmatrix}. \tag{42}$$

We now design a nonlinear controller based on integrator backstepping. If ϕ, θ and Ψ are small ($\sin\theta \approx \theta$ and $\cos\theta \approx \theta$), then $\omega \approx [\dot{\phi}\ \dot{\theta}\ \dot{\Psi}]^T$ and $\dot{\omega} \approx [\ddot{\phi}\ \ddot{\theta}\ \ddot{\Psi}]^T$, and hence the equation of motions can be simplified as

$$\ddot{x} = \theta u_1, \tag{43}$$

$$\ddot{y} = -\phi u_1, \tag{44}$$

$$\ddot{z} = -g + u_1, \tag{45}$$

$$\ddot{\phi} = f_1(\phi, \theta, \Psi) + u_2, \tag{46}$$

$$\ddot{\theta} = f_2(\phi, \theta, \Psi) + u_3, \tag{47}$$

$$\ddot{\Psi} = f_3(\phi, \theta, \Psi) + u_4, \tag{48}$$

where

$$\begin{bmatrix} f_1 \\ f_2 \\ f_3 \end{bmatrix} = \bar{J}^{-1} \left(\tau_g + \begin{bmatrix} J_r \dot{\theta} \Omega_r \\ -J_r \dot{\phi} \Omega_r \\ 0 \end{bmatrix} - (\dot{\bar{J}} + \hat{\omega} \bar{J}) \begin{bmatrix} \dot{\phi} \\ \dot{\theta} \\ \dot{\Psi} \end{bmatrix} \right). \tag{49}$$

3.1. Controller design

In this section a nonlinear controller is designed to stabilize the system (43)–(48) to the desired equilibrium configuration $(x, y, z, \phi, \theta, \Psi) = (x_d, y_d, z_d, \phi_d, \theta_d, \Psi_d)$.

We choose u_1 as

$$u_1 = g - |z - z_d|^a \, sign\,(z - z_d) - |\dot{z}|^b \, sign\,(\dot{z}), \tag{50}$$

where $b \in (0, 1)$, $a > b / (2 - b)$, $i = 1,2$, are controller parameters. The feedback law (50) controls the quadcopter z-dynamics to $(z, \dot{z}) = (z_d, 0)$ in finite time [34] so that $u_1 \to g$ in finite time.

After reaching the desired altitude, Eqs. (43) and (44) take the following form:

$$\ddot{x} = g\theta, \tag{51}$$

$$\ddot{y} = -g\phi. \tag{52}$$

We now apply a backstepping method to design the controls u_2 and u_3 to stabilize the system to the equilibrium at $(x, y, \phi, \theta) = (x_d, y_d, \phi_d, \theta_d)$.

Assume that θ and ϕ are virtual inputs for the x and y subsystems, respectively. Stabilizing feedback functions for the x-subsystem is given by

$$\theta = -k_1(x - x_d) - k_2\dot{x}, \tag{53}$$

$$\phi = k_3(y - y_d) + k_4\dot{y}, \tag{54}$$

where $k_i > 0$, $i = 1,\ldots,4$, so that

$$\ddot{x} + gk_2\dot{x} + gk_1(x - x_d) = 0, \tag{55}$$

$$\ddot{y} + gk_3\dot{y} + gk_4(y - y_d) = 0. \tag{56}$$

Define

$$y_1 = \theta + k_1(x - x_d) + k_2\dot{x}, \tag{57}$$

$$y_2 = \phi - k_3(y - y_d) - k_4\dot{y}, \tag{58}$$

and consider the y_1 and y_2 dynamics given by

$$\dot{y}_1 = \dot{\theta} + k_1\dot{x} + k_2\ddot{x} = \dot{\theta} + k_1\dot{x} + k_2g\theta, \tag{59}$$

$$\dot{y}_2 = \dot{\phi} - k_3\dot{y} - k_4\ddot{y} = \dot{\phi} - k_3\dot{y} + k_4g\phi. \tag{60}$$

Define the sliding variables (s_1, s_2)

$$s_1 = \dot{y}_1 + \alpha_1 y_1, \tag{61}$$

$$s_2 = \dot{y}_2 + \alpha_2 y_2, \tag{62}$$

where $\alpha_i > 0$, $i = 1,2$, which can be simplified as

$$s_1 = \dot{\theta} + (k_2g + \alpha_1)\theta + (k_1 + \alpha_1k_2)\dot{x} + \alpha_1k_1(x - x_d), \tag{63}$$

$$s_2 = \dot{\phi} + (k_4g + \alpha_2)\phi - (k_3 + \alpha_2k_4)\dot{y} - \alpha_2k_3(y - y_d). \tag{64}$$

The dynamics of sliding variables are found simply by taking time derivative of the sliding variables as

$$\dot{s}_1 = \ddot{\theta} + (k_2g + \alpha_1)\dot{\theta} + (k_1 + \alpha_1k_2)g\theta + \alpha_1k_1\dot{x}, \tag{65}$$

$$\dot{s}_2 = \ddot{\phi} + (k_4g + \alpha_2)\dot{\phi} + (k_3 + \alpha_2k_4)g\phi - \alpha_2k_3\dot{y}. \tag{66}$$

Substituting the expressions for $\ddot{\phi}$ and $\ddot{\theta}$ from (46) and (47), respectively, we obtain

$$\dot{s}_1 = f_2(\phi, \theta, \Psi) + u_3 + (k_2g + \alpha_1)\dot{\theta} + (k_1 + \alpha_1k_2)g\theta + \alpha_1k_1\dot{x}, \tag{67}$$

$$\dot{s}_2 = f_1(\phi, \theta, \Psi) + u_2 + (k_4g + \alpha_2)\dot{\phi} + (k_3 + \alpha_2k_4)g\phi - \alpha_2k_3\dot{y}. \tag{68}$$

We choose the inputs u_2 and u_3 as

$$u_2 = -\lambda_2 \,\text{sign}\,(s_2) - f_1(\phi, \theta, \Psi) - (k_4 g + \alpha_2)\dot{\phi} - (k_3 + \alpha_2 k_4) g\phi + \alpha_2 k_3 \dot{y}, \tag{69}$$

$$u_3 = -\lambda_1 \,\text{sign}\,(s_1) - f_2(\phi, \theta, \Psi) - (k_2 g + \alpha_1)\dot{\theta} - (k_1 + \alpha_1 k_2) g\theta - \alpha_1 k_1 \dot{x}, \tag{70}$$

so that the following closed-loop response for the sliding variables is obtained:

$$\dot{s}_1 = -\lambda_1 \,\text{sign}\,(s_1), \tag{71}$$

$$\dot{s}_2 = -\lambda_2 \,\text{sign}\,(s_2), \tag{72}$$

where we choose $\lambda_1 > 0$ and $\lambda_2 > 0$ large enough so that the terms F_d and τ_d are dominated by the sliding mode terms.

Now consider the Ψ-dynamics given by (48). The following control law stabilizes the Ψ-dynamics to $(\Psi, \dot{\Psi}) = (\Psi_d, 0)$:

$$u_4 = -k_5(\Psi - \Psi_d) - k_6\dot{\psi} - f_3(\phi, \theta, \Psi), \tag{73}$$

where $k_5, k_6 > 0$.

The voltage inputs V_i, $i = 1, \ldots, 4$, are determined by substituting the expressions for the virtual control inputs u_i, $i = 1, \ldots, 4$, into Eqs. (34) and (42).

Consider Eqs. (31) and (35), and ignore F_d and τ_d. Then we have

$$\dot{v} = M^{-1}F_t - M^{-1}\hat{\omega}Mv, \tag{74}$$

$$\dot{\omega} = \overline{J}^{-1}\tau_r - \overline{J}^{-1}(\dot{\overline{J}} + \hat{\omega}\overline{J})\omega. \tag{75}$$

Eq. (30) can be rewritten as

$$\ddot{\eta} = \overline{m}^{-1}\left(-B_t^T\dot{v} - B_a^T\dot{w} + \tau_s - \dot{B}_t^T v - \dot{B}_a^T w - \dot{\overline{m}}\dot{\eta} + \frac{\partial L}{\partial \eta}\right), \tag{76}$$

which can be expressed in terms of F_t and τ_r as

$$\ddot{\eta} = \overline{m}^{-1}\left(\tau_s - \frac{B_t^T F_t}{m_t} - B_t^T\hat{\omega}v - B_a^T\overline{J}^{-1}(\tau_r - (\dot{\overline{J}} + \hat{\omega}\overline{J})\omega) - \dot{B}_t^T v - \dot{B}_a^T w - \dot{\overline{m}}\dot{\eta} + \frac{\partial L}{\partial \eta}\right). \tag{77}$$

In order to have exponential convergence of the shape variables η to the desired η_d we choose τ_s as

$$\tau_s = \frac{B_t^T F_t}{m_t} + B_t^T\hat{\omega}v + B_a^T\overline{J}^{-1}(\tau_r - (\dot{\overline{J}} + \hat{\omega}\overline{J})\omega) + \dot{B}_t^T v + \dot{B}_a^T w + \dot{\overline{m}}\dot{\eta} - \frac{\partial L}{\partial \eta} - \overline{m}\left(2\lambda\dot{\eta} + \lambda^2(\eta - \eta_d)\right) \tag{78}$$

where $\lambda > 0$, so that

$$\ddot{\eta} + 2\lambda\dot{\eta} + \lambda^2(\eta - \eta_d) = 0. \tag{79}$$

4. Simulation

The controller developed in the previous sections is applied to the full nonlinear model of the aerial robot. The relevant parameter values of the system are listed in **Table 1**.

A rest-to-rest motion was simulated with initial conditions $(x_0, y_0, z_0) = (0, 0, 0)$, $(\phi_0, \theta_0, \Psi_0) = (0, 0, 0)$, and $(\theta_{10}, \theta_{20}) = (0, 0)$. The desired position, attitude, and joint angles were set as $(x_d, y_d, z_d) = (30, 50, 40)$ [m], $(\phi_d, \theta_d, \Psi_d) = (0, 0, 0)$, and $(\theta_{1d}, \theta_{2d}) = (30, 60)$ [°], respectively.

The control parameters are chosen as

$$(k_1, k_2, k_3, k_4, k_5, k_6) = (2, 0.1, 2, 0.1, 2, 2), \tag{80}$$

$$(\lambda_1, \lambda_2) = (1, 1), \ (\alpha_1, \alpha_2) = (0.1, 0.1). \tag{81}$$

As shown in **Figures 3–5**, the position, attitude, and joint angles converge to their desired values in around 40 s. **Figure 6** shows the time responses of the control inputs u_i, $i = 1,\ldots,4$.

Symbol	Parameter	Value	Unit
K_v	Transformation constant	54.945	rad s V^{-1}
J_r	Rotor inertia	6×10^{-5}	kg m^2
J_{xx}	MOI about x axis	0.0552	kg m^2
J_{yy}	MOI about y axis	0.0552	kg m^2
J_{zz}	MOI about z axis	0.1104	kg m^2
b	Thrust coefficient	3.935139×10^{-6}	N V^{-1}
d	Drag coefficient	1.192564×10^{-7}	Nm V^{-1}
l	Distance from pivot to motor	0.1969	m
m	Mass	2.85	kg
g	Acceleration of gravity	9.81	ms^{-2}
V	Maximum input voltage	10	V
m_1	Mass of link 1	0.1	kg
m_2	Mass of link 2	0.1	kg
m_p	Mass of the payload	0.1	kg
l_1	Length of link 1	0.5	m
l_2	Length of link 2	0.5	m

Table 1. Parameters of the aerial robot.

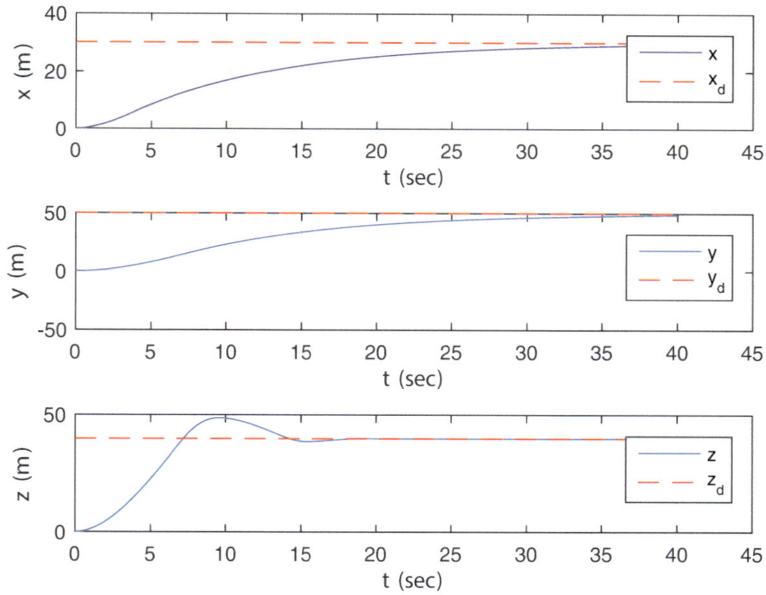

Figure 3. Time responses of the aerial robot's position x, y, and z.

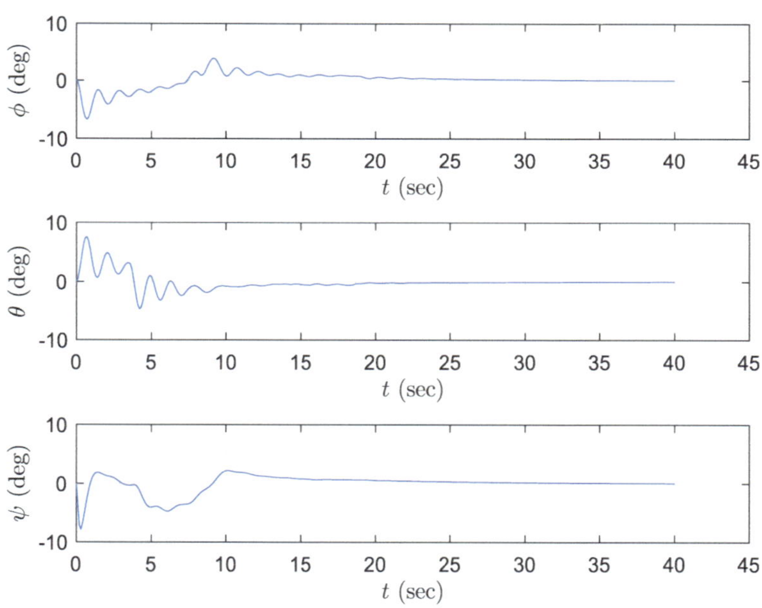

Figure 4. Time responses of the aerial robot's Euler angles ϕ, θ, and Ψ.

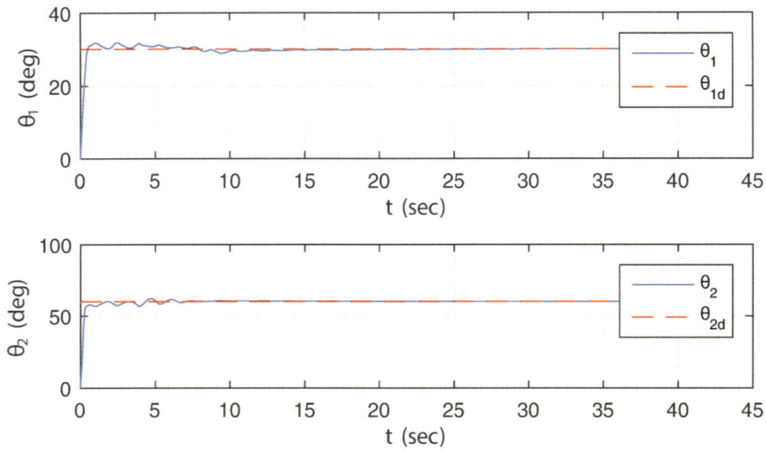

Figure 5. Time responses of the robotic arm's joint angles θ_1 and θ_2.

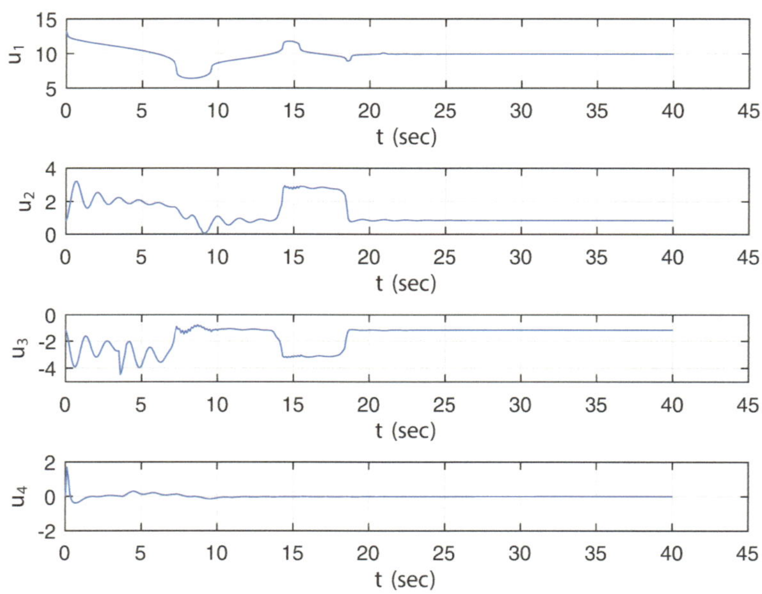

Figure 6. Time responses of the control inputs u_i, $i = 1,\ldots,4$.

5. Conclusions

This chapter first derives the complete nonlinear dynamics of an aerial robot consisting of a quadcopter with a two-link robot manipulator. Precise control of such an aerial robot is a challenging task since the translational and rotational dynamics of the quadcopter are strongly coupled with the dynamics of the manipulator. We extend our previous results on the control of quadrotor UAVs to the control of aerial robots. In particular, we design a backstepping and

Lyapunov-based nonlinear feedback control law that achieves the point-to-point control of the areal robot. The effectiveness of this feedback control law is illustrated through a simulation example.

The many avenues considered for future research include problems involving collaborative control of multiple aerial robots. Future research also includes designing nonlinear control laws that achieve robustness, insensitivity to system and control parameters, and improved disturbance rejection. We also plan to explore the use of geometric mechanics formulation of such control problems.

Appendix A

The matrices M and $\bar{J}$ can be expressed as

$$M = m_t \begin{bmatrix} 1 & 0 & 0 \\ 0 & 1 & 0 \\ 0 & 0 & 1 \end{bmatrix}, \ \bar{J} = \begin{bmatrix} \bar{J}_{11} & \bar{J}_{12} & \bar{J}_{13} \\ \bar{J}_{21} & \bar{J}_{22} & \bar{J}_{23} \\ \bar{J}_{31} & \bar{J}_{32} & \bar{J}_{33} \end{bmatrix},$$

where

$$\begin{aligned}
\bar{J}_{11} &= J_{xx} + [m_1 l_{c1}^2 + (m_2 + m_p) l_1^2 + \frac{1}{12} m_1 l_1^2] \sin^2\theta_1 + [m_2 l_{c2}^2 + m_p l_2^2 + \frac{1}{12} m_2 l_2^2] \sin^2\theta_2 \\
&\quad + 2 l_1 (m_2 l_{c2} + m_p l_2) \sin\theta_1 \sin\theta_2, \\
\bar{J}_{22} &= J_{yy} + m_1 l_{c1}^2 + (m_2 + m_p) l_1^2 + (m_2 l_{c2}^2 + m_p l_2^2) + 2 l_1 (m_2 l_{c2} + m_p l_2) \cos(\theta_2 - \theta_1) \\
&\quad + \frac{1}{12} [m_1 l_1^2 + m_2 l_2^2], \\
\bar{J}_{33} &= J_{zz} + [m_1 l_{c1}^2 + (m_2 + m_p) l_1^2 + \frac{1}{12} m_1 l_1^2] \cos^2\theta_1 + [m_2 l_{c2}^2 + m_p l_2^2 + \frac{1}{12} m_2 l_2^2] \cos^2\theta_2 \\
&\quad + 2 l_1 (m_2 l_{c2} + m_p l_2) \cos\theta_1 \cos\theta_2, \\
\bar{J}_{12} &= \bar{J}_{21} = \bar{J}_{23} = \bar{J}_{32} = 0, \\
\bar{J}_{13} &= \bar{J}_{31} = -\frac{1}{2} [m_1 l_{c1}^2 + (m_2 + m_p) l_1 + \frac{1}{12} m_1 l_1^2] \sin 2\theta_1 \\
&\quad - l_1 (m_2 l_{c2} + m_p l_2) \sin(\theta_1 + \theta_2) - \frac{1}{2} [(m_2 l_{c2}^2 + m_p l_2^2) + \frac{1}{12} m_2 l_2^2] \sin 2\theta_2.
\end{aligned}$$

The matrix $\overline{m}$ can be computed as

$$\overline{m} = \begin{bmatrix} m_1 l_{c1}^2 + (m_2 + m_p) l_1^2 + \frac{1}{12} m_1 l_1^2 & (m_2 l_{c2} + m_p l_2) l_1 \cos(\theta_2 - \theta_1) \\ (m_2 l_{c2} + m_p l_2) l_1 \cos(\theta_2 - \theta_1) & m_2 l_{c2}^2 + m_p l_2^2 + \frac{1}{12} m_2 l_2^2 \end{bmatrix}.$$

The matrices K, B_r, and B_t are given by

$$K = \begin{bmatrix} 0 & -K_{xy} & 0 \\ K_{xy} & 0 & -K_{xz} \\ 0 & K_{xz} & 0 \end{bmatrix}, \; B_r = \begin{bmatrix} 0 & 0 \\ B_{r1} & B_{r2} \\ 0 & 0 \end{bmatrix},$$

where

$$\begin{aligned} K_{xy} &= [m_1 l_{c1} + (m_2 + m_p) l_1] \sin\theta_1 + (m_2 l_{c2} + m_p l_2) \sin\theta_2, \\ K_{xz} &= [m_1 l_{c1} + (m_2 + m_p) l_1] \cos\theta_1 + (m_2 l_{c2} + m_p l_2) \cos\theta_2, \\ B_{r1} &= -[m_1 l_{c1}^2 + (m_2 + m_p) l_1^2 + (m_2 l_{c2} + m_p l_2) l_1 \cos(\theta_2 - \theta_1)] + \frac{1}{12} m_1 l_1^2, \\ B_{r2} &= -[m_2 l_{c2}^2 + m_p l_2^2 + (m_2 l_{c2} + m_p l_2) l_1 \cos(\theta_2 - \theta_1)] + \frac{1}{12} m_2 l_2^2, \end{aligned}$$

and

$$B_t = \begin{bmatrix} -(m_1 l_{c1} + m_2 l_1 + m_p l_1) \sin\theta_1 & -(m_2 l_{c2} + m_p l_2) \sin\theta_2 \\ 0 & 0 \\ (m_1 l_{c1} + m_2 l_1 + m_p l_1) \cos\theta_1 & (m_2 l_{c2} + m_p l_2) \cos\theta_2 \end{bmatrix}.$$

Author details

Mahmut Reyhanoglu* and Muhammad Rehan

*Address all correspondence to: mreyhanoglu@icloud.com

University of North Carolina Asheville, One University Heights, Asheville, NC, USA

References

[1] Brezoescu A, Espinoza T, Castillo P, Lozano R. Adaptive trajectory following for a fixed-wing UAV in presence of crosswind. Journal of Intelligent and Robotic Systems. 2013;69:257–271

[2] Espinoza T, Dzul A, Llama M. Linear and nonlinear controllers applied to fixed-wing UAV. International Journal of Advanced Robotics Systems. 2006;10:1–10

[3] Ferreira HC, Baptista RS, Ishihara JY, Borges GA. Disturbance rejection in a fixed wing UAV using nonlinear H_∞ state feedback. In: Proceedings of International Conference on Control and Automation. 2011. pp. 386–391

[4] Gavilan F, Acosta JA, Vazquez R. Control of the longitudinal flight dynamics of an UAV using adaptive backstepping. In: Proceedings of the 18th IFAC World Congress. 2011. pp. 1892–1897

[5] Hervas JR, Kayacan E, Reyhanoglu M, Tang H. Sliding mode control of fixed-wing UAVs in the presence of stochastic wind. In: Proceedings of International Conference on Control, Automation, Robotics and Vision. 2014. pp. 986–991

[6] Hervas JR, Reyhanoglu M, Tang H. Automatic landing control of unmanned aerial vehicles on moving platforms. In: Proceedings of IEEE International Symposium on Industrial Electronics. 2014. pp. 69–74

[7] Hervas JR, Reyhanoglu M, Tang H. Nonlinear automatic landing control of unmanned aerial vehicles on moving platforms via a 3D laser radar. In: AIP Proceedings. 2014. pp. 907–917

[8] Hervas JR, Reyhanoglu M, Tang H, Kayacan E. Nonlinear control of fixed-wing UAVs in presence of stochastic winds. Communications in Nonlinear Science and Numerical Simulation. 2016;33:57–69

[9] Kannan S, Alma M, Olivares-Mendez MA, Voos H. Adaptive control of aerial manipulation vehicle. In: Proceedings of IEEE International Conference on Control System, Computing and Engineering. 2014. pp. 1–6

[10] Kayacan E, Khanesar MA, Hervas JR, Reyhanoglu M. Intelligent control of unmanned aerial vehicles using fuzzy neural networks. International Journal of Aerospace Engineering. 2017:2017:1–13

[11] Liu C, McAree O, Chen WH. Path-following control for small fixed-wing unmanned aerial vehicles under wind disturbances. International Journal of Robust and Nonlinear Control. 2013;23:1682–1698

[12] MacKunis W, Wilcox ZD, Kaiser MK, Dixon WE. Global adaptive output feedback tracking control of an unmanned aerial vehicle. IEEE Transactions on Control Systems Technology. 2010;18:1390–1397

[13] Chen X, Wang L. Cascaded model predictive control of a quadrotor UAV. In: Proceedings of Australian Control Conference. 2013. pp. 354–359

[14] Huang H, Hoffmann GM, Waslander SL, Tomlin CJ. Aerodynamics and control of autonomous quadrotor helicopters in aggressive maneuvering. In: Proceedings of International Conference on Robotics and Automation. 2009. pp. 3277–3282

[15] Shulong Z, Honglei A, Daibing Z, Lincheng S. A new feedback linearization LQR control for attitude of quadrotor. In: Proceedings of International Conference on Control, Automation, Robotics and Vision. 2013. pp. 1593–1597

[16] Damen R, Reyhanoglu M, MacKunis W, Hervas JR. Passivity-based quaternion feedback control of a hover system. In: Proceedings of International Conference on Control, Automation and Systems. 2016. pp. 201–206

[17] Reyhanoglu M, Damen R, MacKunis W. Observer-based sliding mode control of a 3-DOF hover system. In: Proceedings of International Conference on Control, Automation, Robotics and Vision. 2016. pp. 1–6

[18] Stebler S, Campobasso M, Kidambi K, MacKunis W, Reyhanoglu M. Dynamic neural network-based sliding mode estimation of quadrotor systems. In: Proceedings of American Control Conference. 2017. pp. 1–6

[19] Stebler S, MacKunis W, Reyhanoglu M. Nonlinear output feedback tracking control of a quadrotor UAV in the presence of uncertainty. In: Proceedings of International Conference on Control, Automation, Robotics and Vision. 2016. pp. 1–6

[20] Alvarez-Munoz JU, Marchand N, Guerrero-Castellanos F, Durand S, Lopez-Luna AE. Improving control of quadrotors carrying a manipulator arm. In: Proceedings of XVI Congreso Latinoamericano de Control Automatico. 2014. pp. 1–6

[21] Arleo G, Caccavale F, Muscio G, Pierri F. Control of quadrotor aerial vehicles equipped with a robotic arm. In: Proceedings of Mediterranean Conference on Control & Automation. 2013. pp. 1–7

[22] Heredia G, Jimenez-Cano AE, Sanchez I, Llorente D, Vega V, Braga J, Acosta JA, Ollero A. Control of a multirotor outdoor aerial manipulator. In: Proceedings of IEEE/RSJ International Conference on Intelligent Robots and Systems. 2014. pp. 3417–3422

[23] Kang Y, Hedrick JK. Linear tracking for a fixed-wing UAV using nonlinear model predictive control. IEEE Transactions on Control Systems Technology. 2009;17:1202–1210

[24] Kobilarov M. Nonlinear trajectory control of multi-body aerial manipulators. Journal of Intelligent Robotic Systems. 2014;73:679–692

[25] Lee D, Nataraj C, Burg TC, Dawson DM. Adaptive tracking control of an underactuated aerial vehicle. In: Proceedings of American Control Conference. 2011. pp. 2326–2331

[26] Lee D, Burg TC, Dawson DM, Shu D, Xian B, Tatlicioglu E. Robust tracking control of an underactuated quadrotor aerial-robot based on a parametric uncertain model. In: Proceedings of IEEE International Conference on Systems, Man and Cybernetics. 2009. pp. 3187–3192

[27] Orsag M, Korpela C, Pekala M, Oh P. Stability control in aerial manipulation. In: Proceedings of American Control Conference. 2013. pp. 5581–5586

[28] Cho S, McClamroch NH, Reyhanoglu M. Dynamics of multibody vehicles and their formulation as nonlinear control systems. In: Proceedings of American Control Conference. 2000. pp. 3908–3912

[29] Meirovitch L, Kwak MK. State equations for a spacecraft with maneuvering flexible appendages in terms of quasi-coordinates. Applied Mechanics Reviews. 1989;42:161–170

[30] Reyhanoglu M, van der Schaft AJ, McClamroch NH, Kolmanovsky I. Nonlinear control of a class of underactuated systems. Proceedings of IEEE Conference on Decision and Control. 1996. pp. 1682–1687

[31] Reyhanoglu M, van der Schaft AJ, McClamroch NH, Kolmanovsky I. Dynamics and control of a class of underactuated mechanical systems. IEEE Transactions on Automatic Control. 1999;44:1663–1671

[32] Reyhanoglu M, Cho S, McClamroch NH. Discontinuous feedback control of a special class of underactuated mechanical systems. International Journal of Robust and Nonlinear Control. 2000;10:265–281

[33] Reyhanoglu M. Maneuvering control problems for a spacecraft with unactuated fuel slosh dynamics. In: Proceedings of IEEE Conference on Control Applications. 2003. pp. 1–6

[34] Haimo VT. Finite time controllers. SIAM Journal on Control and Optimization. 1986;24: 760–770

Chapter 6

Design and Development of Aerial Robotic Systems for Sampling Operations in Industrial Environment

Rodrigo Munguia, Edmundo Guerra, Sarquis Urzua,
Yolanda Bolea and Antoni Grau

Additional information is available at the end of the chapter

http://dx.doi.org/10.5772/intechopen.70005

Abstract

This chapter describes the development of an autonomous fluid sampling system for outdoor facilities, and the localization solution to be used. The automated sampling system will be based on collaborative robotics, with a team of a UAV and a UGV platform travelling through a plant to collect water samples. The architecture of the system is described, as well as the hardware present in the UAV and the different software frameworks used. A visual simultaneous localization and mapping (SLAM) technique is proposed to deal with the localization problem, based on authors' previous works, including several innovations: a new method to initialize the scale using unreliable global positioning system (GPS) measurements, integration of attitude and heading reference system (AHRS) measurements into the recursive state estimation, and a new technique to track features during the delayed feature initialization process. These procedures greatly enhance the robustness and usability of the SLAM technique as they remove the requirement of assisted scale initialization, and they reduce the computational effort to initialize features. To conclude, results from experiments performed with simulated data and real data captured with a prototype UAV are presented and discussed.

Keywords: UAV, collaborative robotics, inertio-visual odometry, visual SLAM, GPS

1. Introduction

The development of aerial robots has become one of the most active fields of research in the last decade. Innovations in multiple fields, like the lithium polymer batteries, microelectromechanical sensors, more powerful propellers, and the availability of new materials and prototyping technologies, have opened the field to researchers and institutions, who used to be denied

INTECH
open science | open minds

© 2017 The Author(s). Licensee InTech. This chapter is distributed under the terms of the Creative Commons Attribution License (http://creativecommons.org/licenses/by/3.0), which permits unrestricted use, distribution, and reproduction in any medium, provided the original work is properly cited. (cc) BY

access given the costs, both economic and in specialized personnel. The new-found popularity of this research field has led to the proliferation of advancements in several areas [1] which used to ignore the possibilities of aerial robots given the limited capacities they presented.

One of the environments where aerial robots are putting a foothold for the first time is industry. Given the level of accountability, certification, and responsibility required in industry, the field had been always reluctant to the introduction of experimental state of the art technologies. But thanks to the wider experimentation in aerial robots, resilience and performance robustness levels have been improving, making them an option when solving several industry problems. Currently, these problem are focused in logistical aspects and related operations, like distribution of good and placement of them in otherwise hard to access points. An example of this kind of applications would be the surveying and monitoring of fluids in installations with multiple basins and tanks, for example, a wastewater processing plant.

Basin sampling operations generally require multiple samples of fluids in several points with a given periodicity. This makes the task cumbersome, repetitive, and depending on the features of the environment and other factors, potentially dangerous. As such, automatization of the task would provide great benefits, reducing the efforts and risks taken by human personnel, and opening options in terms of surveying scheduling.

One of the challenges that any autonomous Unmanned Aerial Vehicle (UAV) has to face is that of estimating its pose with respect to the relevant navigational frames with accuracy. The estimation methodology here discussed was formulated for estimating the state of the aerial vehicle. In this case, the state is composed of the variables defining the location and attitude as well as their first derivatives. The visual features seen by the camera are also included into the system state. On the other hand, the orientation estimation can be estimated in a robust manner by most flight management units (FMUs), with the output of the attitude and heading reference system (AHRS) frequently used as a feedback to the control system for stabilization.

In order to account for the uncertainties associated with the estimation provided by the attitude and heading reference systems (AHRSs), the orientation is included into the state vector and is explicitly fused into the system. Regarding the problem of position estimation, it cannot be solved for applications that require performing precise maneuvers, even with global positioning system (GPS) signal available. Therefore, some additional sensory information is integrated into the system in order to improve its accuracy, namely monocular vision.

The use of a monocular camera as the unique sensory input of a simultaneous localization and mapping (SLAM) system comes with a difficulty: the estimation of the robot trajectory, as well as the features map, can be carried out without metric information. This problem was pointed out since early approaches like [2]. If the metric scale wants to be recovered, it is necessary to incorporate some source of metric information into the system. In this case, the GPS and the monocular vision can operate in a complementary manner. In the proposed method, the noisy GPS data are used to incorporate metric information into the system, in periods where it is available. On the other side, the monocular vision is used for refining the estimations when the GPS is available or for performing purely visual-based navigation in periods where the GPS is unavailable.

In this work, we present an automated system designed with the goal of automate sampling tasks in an open air plant and propose a solution for the localization problem in industrial environments when GPS data are unreliable. The final system should use two autonomous vehicles: a robotic ground platform and a UAV, which would collaborate to collect batches of samples in several tanks. The specifications and designs of the system are described, focusing on the architecture and the UAV. The next section describes the vision-based solution proposed to deal with the localization for navigation problem, commenting several contributions done with respect to a classical visual SLAM approach. Results discussing the performance and accuracy of said localization techniques are presented, both based on simulations and real data captured with the UAV described in Section 2. Finally, the conclusions discuss the next step in the testing and development of the system and the refining of the localization technique.

2. System architecture

The architecture proposed to deal with the fluid sampling task aims to maximize the capability to reach with accuracy the desired points of operation and measurement; and minimize the risks associated with the process. The risks for human operators are removed or minimized, as they can perform their tasks without exposing themselves to the outdoor industrial environment. To achieve this, the system will present two different robotic platforms: a quadcopter UAV acts as sample collector, picking fluid samples from the tanks; and a Unmanned Ground Vehicle (UGV) platform acts as a collector carrier, transporting both the collector and the samples. As the risks associated with the operation of the sample collector UAV are mainly related to the flight operations, the UAV will travel generally safely landed on the collector carrier, where it could be automatically serviced with replacement sample containers or battery charges.

2.1. Sampling system architecture and communications

The designed architecture of the system and its expected operation process can be observed in **Figure 1**. The architecture has been divided into several blocks, so it can fit into a classical deployment scheme in an industrial production environment. The analytics technicians at the laboratory can use the scheduling and control module interface to order the collection of a batch of samples. This is called collection order, detailing a number of samples, which basin must they come from, and if there are any required sampling patterns or preferences. The process can be set to start at a scheduled time, and stops may be enforced, i.e., samples on a specific tank cannot be taken before a set hour.

The collection order is formed by a list of GPS coordinates (with optional parameters, like times to perform operations), which is processed in the central server of the system. This produces a mission path, which includes a route for the sample carrier, with one or more stops. The mission path also includes the sampling flight that the sample collector UAV has to perform with the sample carrier stop. The data of this flight, called collection mission, will be transmitted by the sample carrier to the sample collector and will contain a simple trajectory that the sample collector has to approximately follow, with height indications to avoid obstacles, until the sampling point is reached.

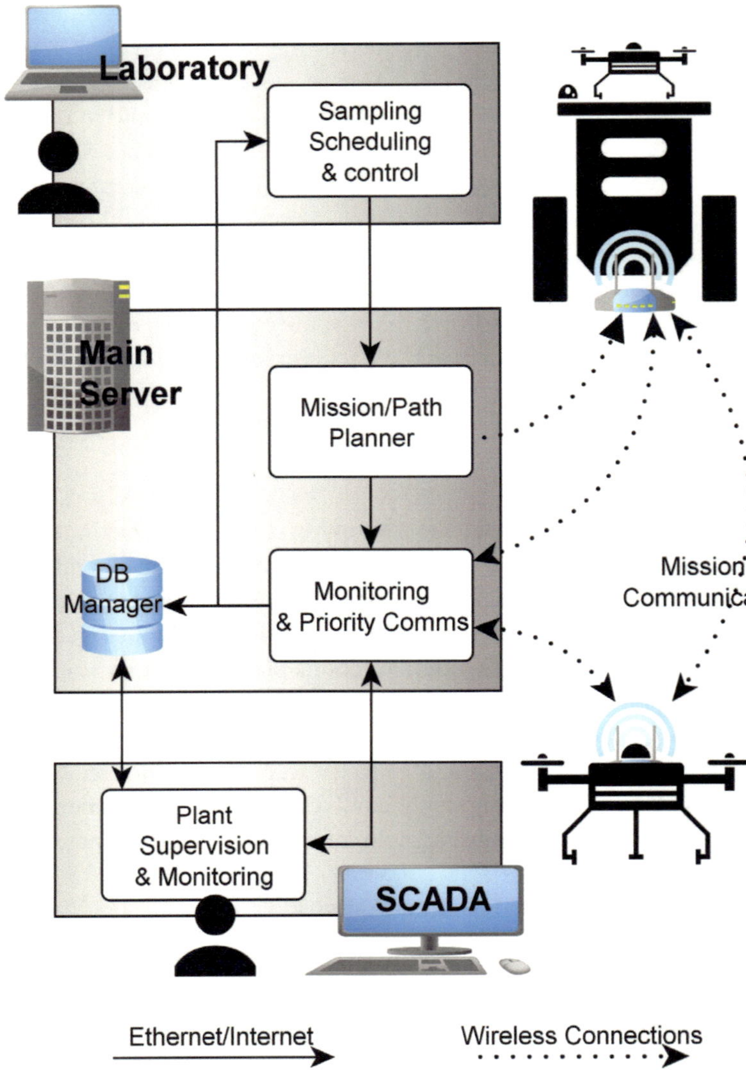

Figure 1. Proposed system distribution and communications diagram.

The different paths and trajectories are generated by the path planner module, which works on a two-dimensional (2D) grid map model of the outdoors. When data are available, the grid is to be textured with the map obtained from the visual SLAM technique proposed. The utilization of the grid model allows introducing additional data, like occupancy, possible obstacles, or even scheduling the accessibility of an area, e.g., we can set a path normally used by workers during certain hours to be avoided at those times. An energy minimization planning technique, essentially a simplified approach to Ref. [3], is used to obtain the trajectories, considering a set of criteria.

1. All samples in the same basin/tank should be capture consecutively;
2. Minimizing the expected flight effort for the UAV (heuristically assuming that the trajectory is a polyline of vertical and horizontal movements);

3. Minimize penalties for restricted access areas;
4. Minimize the trajectory of the UGV.

The communications with the autonomous platforms of the system will be performed through 4G in order to allow video streaming during the prototyping and testing phases. The data produced by the different elements of the system will be stored and logged to allow posterior analysis. The rest of the communications can be performed through any usual channel in industrial environments, be it a local network, Internet, VPN…, as the segmented architecture allows discrete deployment. The communications between the sample carrier and the sample collector will be performed through ZigBee, although they present Wi-Fi modules to ease development and maintenance tasks.

A subset of the communication protocols is considered priority communications. This includes the supervision and surveying messages in normal planned operation, and those signals and procedures that can affect or override normal operation, e.g., emergency recall or landing. For the UAV, the recall protocol uses a prioritized list of possible fallback points so that the UAV tries to reach them. If the sample carrier is present, the UAV will try to land on it, searching for its landing area through fiduciary markers. If it is not possible, the UAV will try another fallback point in the list, until it lands or the battery is below a certain threshold: in that case, it will just land in the first clear patch of ground available, though this will generally require dropping the sampling device.

2.2. Sample collector aerial robot architecture

The UAV built as a sample collector, UAV is a 0.96 m diameter quadcopter deploying four 16″ propellers with T-Motor MN4014 actuators. The custom-built frame supports the propeller blocks at 5° angle and is made of aluminum and carbon fiber. A PIXHAWK kit is used as a flight management unit (FMU), with custom electronics to support 2 6 s 8000 mAh batteries. An Odroid U4 single-board-computer is used to perform the high level task and deal with all noncritical processes. Beyond the sensors present in the FMU (which includes AHRS and GPS), the UAV presents a front facing camera USB (640×480@30fps) for monitoring purposes, an optical flow with ultrasound sensor facing downward, and a set of four ultrasound sensors deployed in a planar configuration to detect obstacles collision. The FX4 stack [4] is used to manage flying and navigation, while a Robot Operating System (ROS) [5] distribution for ARM architectures is run at the Odroid Single Board Computer (SBC), supporting MavLink for communications. The communication modules and sensors are described at hardware level in **Figure 2**.

The hardware used weights approximately 4250 kg, while the propellers provide a theoretical maximum lift of 13,900 g. The sample collection device, including the container, is being designed with a weight below of 300 g, meaning that UAV, with the sample capture system and up to 1000 ml of water-like fluid, would keep the weight/lift ratio around 0.4.

A simplified diagram of the operation process is shown in **Figure 3**. The communications are divided into two blocks, those connecting the UAV with the main network for supervision and emergency control, including video streaming with MavLink [6] over 4G, and those that will connect it to the sample carrier during routine operation. This routine operation includes receiving

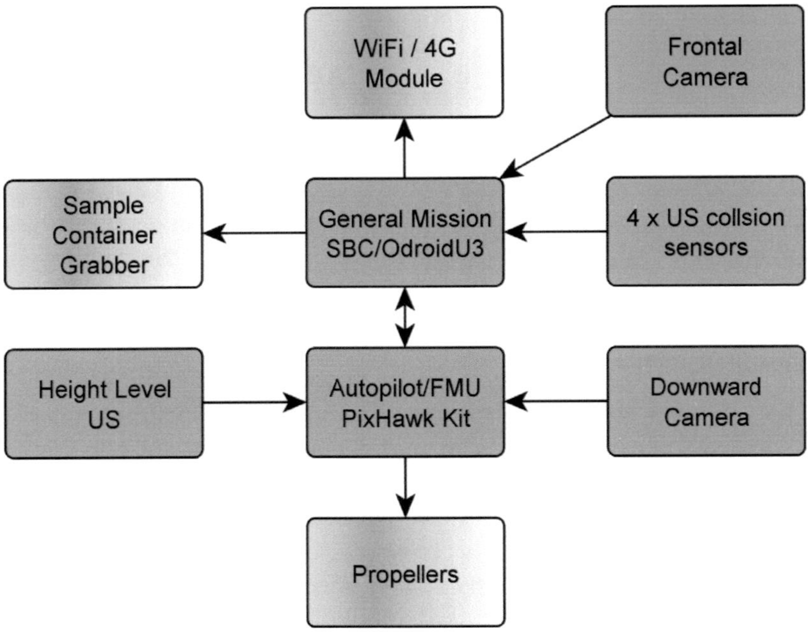

Figure 2. Sample collector aerial robot hardware and communications diagram.

the collection mission detailing the trajectory to the sampling point and the signal to start the process. The action planner module in the UAV supervises the navigation tasks, making sure that the trajectory waypoints are reached and the managing the water sample collector control module.

The localization and positioning of the UAV are solved through a combination of GPS and visual odometry estimation (see Section 4). To approach the water surface, the downward looking ultrasonic height sensor will be used, as vision-based approaches are unreliable on reflective surfaces. To ease the landing problem, the sample carrier will present fiduciary markers to estimate the pose of the drone. This allows landing operations while inserting the sample container into a socket, similarly to an assisted peg-in-the-hole operation. After the UAV is landed, the sample collector control module releases the sample container, and the sample carrier will replace the container with a clean one. After the container is replaced, the carrier will send a signal to the UAV so that the new container is properly locked, while the filled one will be stored.

As the autonomous navigation depends mainly on the localization and positioning, it is largely based in GPS and visual odometry. This combination allows, combined with the ultrasound to find height and avoid possible obstacle, to navigate the environment event if the GPS signal is unreliable. Though there are more accurate alternatives to solve the SLAM problem in terms of sensors, namely using RGB-D cameras or Lighting Detection and Ranging (LiDAR)-based approaches, they present several limitations that make them unsuitable for outdoor industrial environments. These limitations are in addition to the penalty imposed by their economical cost, especially for models industry levels of performance and reliability.

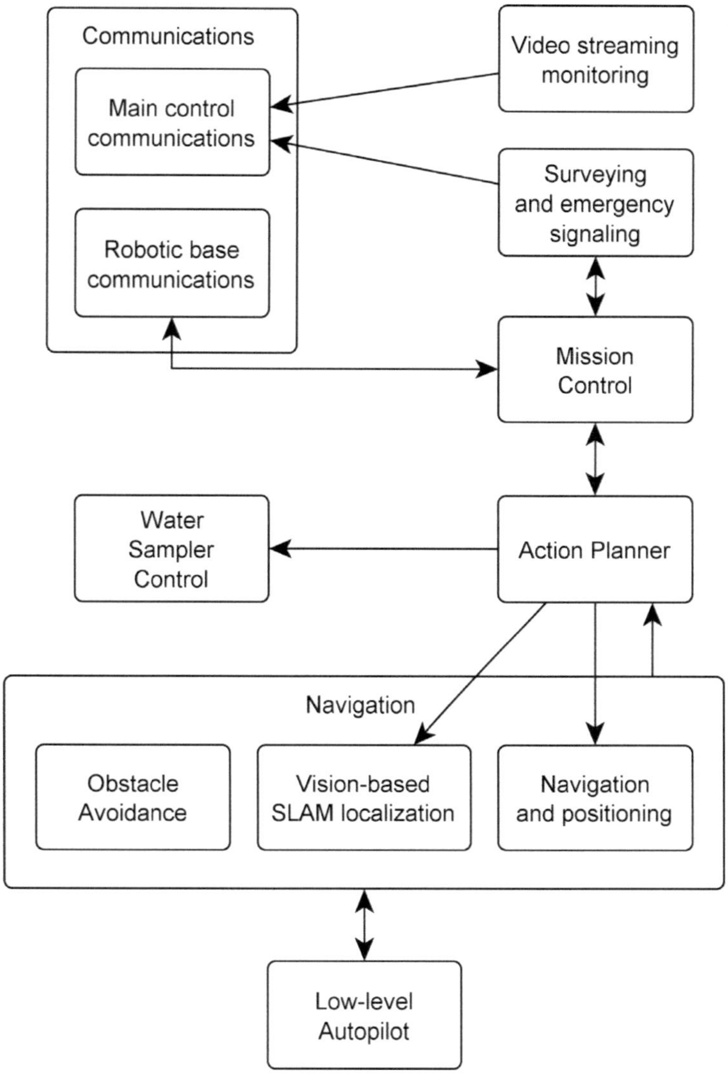

Figure 3. Sample collector UAV robot block diagram.

In the case of RGB-D, most of the sensors found in the market are unreliable in outdoors as they use IR or similar lightning frequencies. The subset of time-of-flight RGB-D sensors presents the same limitation as the LiDAR sensors: they are prone to spurious measurements in environments where the air is not clear (presence of dusty, pollen and particles, vapors from tanks), they present large latencies, making them unfit for real-time operation of UAVs, and they are generally considered not robust enough for industrial operation. In the case of LiDAR, and RGB-D if the SLAM/localization technique used focuses on depth measurements, there is an additional issue present: these approaches normally rely in computationally demanding optimization techniques to achieve accurate results, and the computer power available in an UAV is limited.

3. UAV localization and visual odometry estimation

The drone platform is considered to freely move in any direction in R3 × SO(3)R3 × SO(3), as shown in **Figure 4**.

As the proposed system is mainly intended for local autonomous vehicle navigation, i.e., localize the sample collector during the different collection missions, the local tangent frame is used as the navigation reference frame. The initial position of the sample collector landed over the sample carrier is used to define the origin of the navigation coordinates frame, and the axes are oriented following the navigation convention NED (North, East, Down). The magnitudes expressed in the navigation, sample collector drone (robot), and camera frame are denoted, respectively, by the superscripts NN, RR, and CC. All the coordinate systems are right-handed defined. The proposed method will be taken mainly into account the AHRS and the downward monocular camera, though it also uses data from the GPS during an initialization step.

The monocular camera is assumed to follow the central-projection camera model, with the image plane in front of the origin, thus forming a noninverted image. The camera frame C is considered right handed with the z-axis pointing toward the field of view. It is considered that

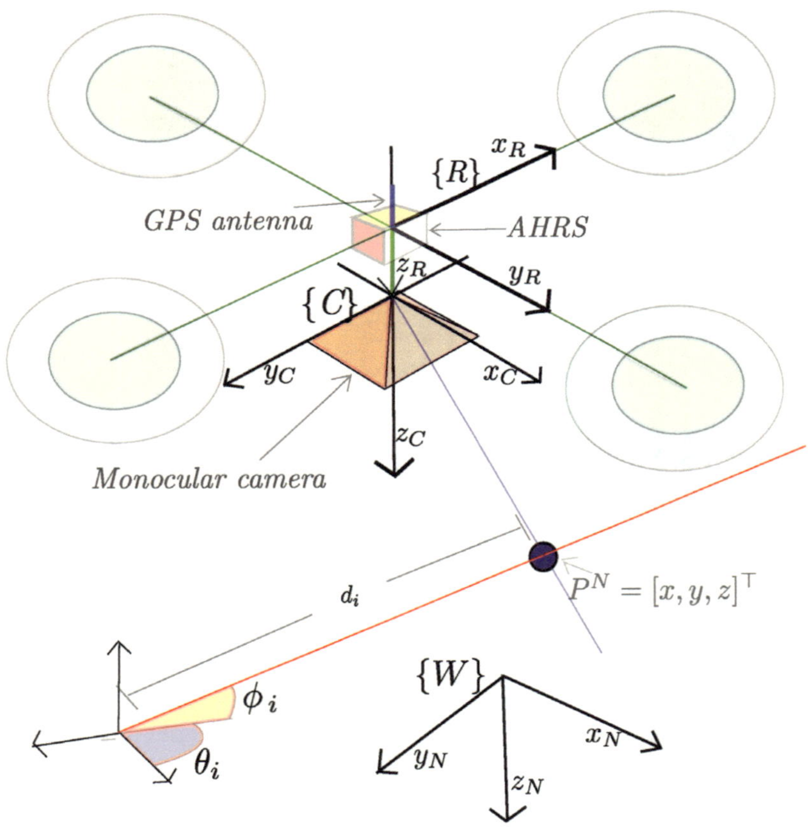

Figure 4. Coordinate systems: the local tangent frame is used as the navigation reference frame NN. AHRS: attitude and heading reference system.

the pixel coordinates are denoted with the $[u, v]$ pair convention and follow the classical direct and inverse observation models [7].

The attitude and heading reference system (AHRS) is a device used for estimating the vehicle orientation, while it is maneuvering. The most common sensors integrated with AHRS devices are gyroscopes, accelerometers, and magnetometers. The advances in micro-electro-mechanical systems (MEMS) and microcontrollers have contributed to the development of inexpensive and robust AHRS devices (e.g., [8–10]).

In the case of the deployed FMU, the accuracy and reliability provided by its AHRS are enough to directly fuse its data into the estimation system. Thus, AHRS measurements are assumed to be available at high rates (50–200 Hz) and modeled according to

$$y_a^N = a^N + v_a \tag{1}$$

where $a^N = [\varphi_v, \theta_v, \psi_v]^T$, being φ_v, θ_v, and ψ_v Euler angles denoting, respectively, the roll, pitch, and yaw of the vehicle; with v_a being Gaussian white noise.

The global positioning system (GPS) is a satellite-based navigation system that provides 3D position information for objects on or near the Earth's surface, studied in several works [11, 12]. The user-equivalent range error (UERE) is a measurement of the cumulative error in GPS position measurements caused by multiple sources of error. These error sources can be modeled as a combination of random noise and slowly varying biases [11]. According to a study [13], the UERE is around 4.0 m (σ); in this case, 0.4 m (σ) corresponds to random noise.

In this work, it is assumed that position measurements y_r can be obtained from the GPS unit, at least at the beginning of the trajectory, and they are modeled by

$$y_r = r^N + v_r \tag{2}$$

where v_r is Gaussian white noise and r^N is the position of the vehicle. As GPS measurements are usually in geodetic coordinates, Eq. (2) assumes that they have been converted to the corresponding local tangent frame for navigation, accounting for the transformation between the robot collector frame and the antenna.

3.1. Problem formulation

The objective of the estimation method is to compute the system state x

$$x = \left[x_v, y_1^N, y_2^N \dots, y_n^N\right]^T \tag{3}$$

where the system state x can be divided into two parts: one defined by x_v and corresponding to the sample collector UAV state, and the other one corresponding the map features. In this case, $y^N{}_i$ defines the position of the ith feature map. The UAV state x_v is composed, at the same time, of

$$x_v = \left[q^{NR}, \omega^R, r^N, v^N\right]^T \tag{4}$$

where q^{NR} represents the orientation of the vehicle with respect to the local world (navigation) frame by a unit quaternion, and $\omega^R = [\omega_x, \omega_y, \omega_z]$ represents the angular velocity of the UAV expressed in the same frame of reference. $r^N = [p_x, p_y, p_z]$ represents the position of the UAV expressed in the same navigation frame, with $v^N = [v_x, v_y, v_z]$ denoting the linear velocities. The location of a feature y_i^N (noted as y_i for simplicity) is parameterized in its Euclidean form:

$$y_i^N = \left[p_{x_i}, p_{y_i}, p_{z_i}\right]^T \tag{5}$$

The proposed system follows the classical loop of prediction-update steps in the extended Kalman filter (EKF) in its direct configuration; working at the frequency of the AHRS. Thus, both the vehicle state and the feature estimations are propagated by the filter, see **Figure 5**.

At the start of a prediction-update loop, the collector UAV state estimation x_v takes a step forward through the following unconstrained constant-acceleration (discrete) model:

$$\begin{aligned} q_{k+1}^{NC} &= \left(\cos\|W\| I_{4\times 4} + \frac{\sin\|W\|}{\|W\|} W\right) q_k^{NC} \\ \omega_{k+1}^C &= \omega_k^C + \Omega^C \\ r_{k+1}^N &= r_k^N + v_k^N \Delta t \\ v_{k+1}^N &= v_k^N + V^N \end{aligned} \tag{6}$$

In the model represented by Eq. (6), a closed form solution of $\dot{q} = 1/2(W)q$ is used to integrate the current velocity rotation ω^C over the quaternion q^{NC}.

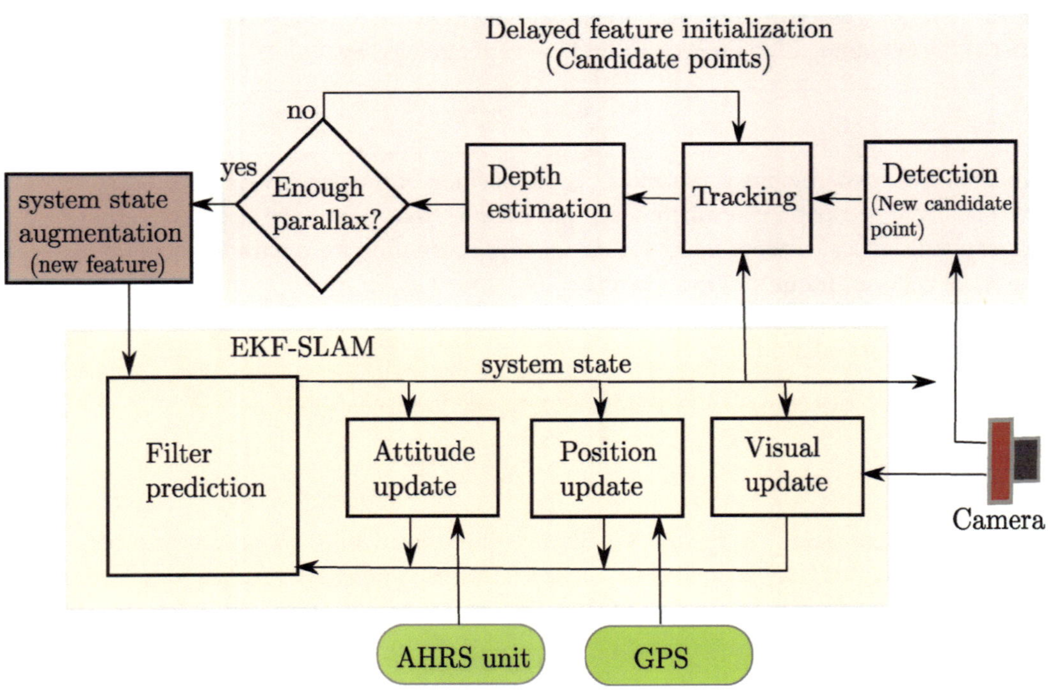

Figure 5. Block diagram showing the architecture of the system. EKF-SLAM: extended Kalman filter simultaneous localization and mapping.

At every step, it is assumed that there is an unknown linear and angular velocity with acceleration zero-mean and known-covariance Gaussian processes σ_v and σ_ω, producing an impulse of linear and angular velocity: $V^N = {\sigma_v}^2\Delta t$ and $\Omega^C = {\sigma_\omega}^2\Delta t$. It is assumed that the map features y_i remain static (rigid scene assumption). Then, the state covariance matrix P takes a step forward by

$$P_{k+1} = \nabla F_x P_k \nabla F_x^T + \nabla F_u Q \nabla F_u^T \tag{7}$$

where Q is the noise covariance matrix of the process (a diagonal matrix with the position and orientation uncertainties); ∇F_x is the Jacobian of the nonlinear prediction model (Eq. (7)); and ∇F_u is the Jacobian of the concerning the derivatives of the input process represented by the unknown linear and angular velocity impulses assumed.

3.2. Visual features: detection, tracking, and initialization

In order to retrieve the depth of a visual feature, the monocular camera, equipped in the UAV, must observe it at least two times while moves along the flight trajectory. The parallax angle is defined by the two projections of those measurements. The convergence of the depth of the features depends on the evolution of the parallax angle. In this work, a method that is based on a stochastic technique of triangulation is used for computing an initial hypothesis of depth for each new feature prior to its initialization into the map. The initialization method is based on previous author's work [7].

3.2.1. Detection stage

The visual-based navigation method requires a minimum number of visual features yi observed at each frame. If this number of visual features is lower than a threshold, then new visual features are initialized into the stochastic map. The Shi-Tomasi corner detector [14] is used for detecting new visual points in the image that will be taken as candidates to be initialized into the map as new features.

When a feature is detected for the first time at k frame, a candidate feature c_l is stored:

$$c_1 = \left[\left(t_{c0}^N \right)^T, \theta_0, \phi_0, P_{y_{ci}}, z_{uv} \right] \tag{8}$$

where $z_{uv} = [u, v]$ is the locationof the visual feature in pixel coordinates in image frame k; $y_{ci} = [{t^N}_{c0}, \theta_0, \phi_0]^T$, complying the inverse observation model $y_{ci} = h(\hat{x}, z_{uv})$. Thus, y_{ci} models a 3D ray originating in the optical center of the camera when the feature is first observed, and pointing to the infinite, with azimuth and elevation θ_0 and ϕ_0, respectively, according to previous work [7]. At the same time, P_{yci} stores, as a 5×5 covariance matrix, the uncertainties of y_{ci}. P_{yci} are computed as $P_{yci} = \nabla Y_{ci} P \nabla {Y_{ci}}^T$, where P is the system covariance and is the Jacobian of the observation model $h(\hat{x}, z_{uv})$ with respect to the state and coordinates u,v. A square patch around $[u, v]$, generally with 11 pixels sides, is also stored, to keep the appearance of the landmark.

3.2.2. Tracking of candidate features

The active search visual technique [15] can be used for tracking visual features that form part of the system state. In this case, the information included in the system state and the system covariance matrix is used for defining images regions where the visual feature is searched. On the other hand, in the case of the candidate features, there is no enough information for applying the active search visual technique. A possibility for tracking the candidate features is to use of general-purpose visual tracking approach [14]. This kind of methods uses only the visual input and do not need information about the system dynamics; however, their computational cost is commonly high. In order to improve the computational performance, a different technique is proposed. The idea is to define very thin elliptical regions of search in the image that are computed based on the epipolar constraints.

After the k frame when the candidate feature was first detected, this one is tracked at subsequent frames $k+1, k+2 \ldots k+n$. In this case, the candidate feature is predicted to lie inside the elliptical region S_c (see **Figure 6**). The elliptical region S_c is centered in the point $[u, v]$, and it is aligned (the major axis) with the epipolar line defined by image points e_1 and e_2.

The epipole is computed by projecting the line $t^N{}_{c0}$ stored in c_l to the current image plane using the camera projective model. As there is not any available depth information, the semiline stored in c_l is considered to have origin $t^N{}_{c0}$, and a length $d = 1$, producing point e_1 and e_2.

The orientation of the ellipse S_c is determined by $\alpha_c = \text{atan2}(e_y, e_x)$, where e_y and e_x represent the y and x coordinates, respectively, of the epipole e, and $e = e_2 - e_1$. The size of the ellipse S_c is determined by its major and minor axis, respectively, a and b. This ellipse S_c represents a probability region where the candidate point must lie in the current frame. The proposed tracking method is intended to be used during an initial short period of time, applying cross-correlation operators. During this period, some information will be gathered in order to compute a depth hypothesis for each candidate point, prior to its initialization as a new map feature.

On the other hand, during this stage, the available information of depth about the candidate features is not well-conditioned. For this reason, it is not easy to determine an adaptive and optimal size of the search image region. Therefore, according to the kind of application, the

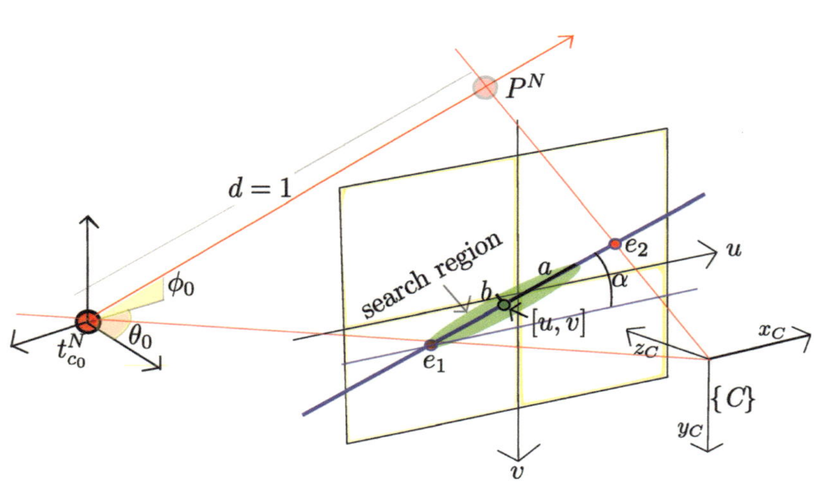

Figure 6. The established search region to match candidate points is constrained to ellipses aligned with the epipolar line.

parameter a is chosen empirically. In the experiments presented in this work, good results were obtained choosing $a = 20$ pixels.

3.2.3. Depth estimation

Each time a new image location $z_{uv}= [u, v]$ is obtained for a given candidate c_l, a depth hypothesis is computed, through stochastic triangulation, as described in previous work [16]. In previous authors' work [17], it was found that the estimates of the feature depth can be improved by passing the hypotheses of depth by a low-pass filter. Thus, only when parallax α_i is greater than a specific threshold ($\alpha_i > \alpha_{min}$), a new feature $y_{new}= [p_{xi}, p_{yi}, p_{zi}]^T= h(c_l, d)$ is added to the system state vector x, as Eq. (3).

In order to correctly operate the proposed method, a minimum number of map features must be maintained inside the camera field of view. For example in Ref. [21], it is stated that a minimum number of three features are required for the operation of monocular SLAM. In practice, of course, it is better to operate with a higher minimum number of features. The above requires to initialize continuously new features into the map.

The initialization time period that takes a candidate point to become a map feature depends directly on the evolution of the parallax angle. At the same time, the evolution of the parallax depends on the depth of the feature and the movement of the camera. For example, the parallax for a near point can increase very quickly due to a small movement of the camera. In practice, it is assumed that the dynamics of the aerial vehicle allows tracking a sufficient number of visual features for initialization and measurement purposes. Experimentally, at least for nonaggressive typical flight maneuvers, we have not found yet major problems in order to accomplish the above requirement.

3.3. Prediction-update loop and map management

Once initialized, the visual features y_i are tracked by means of an active search technique, using zero-mean normalized cross-correlation to match a given feature, and the patch that notes its appearance to a given pixel to be found in a search area. This search area is defined using the innovation covariance. The general methodology for the visual update step can be found in previous works [7, 16, 18], where the details in terms of mathematical representation and implementation are discussed. In this work, the loop closing problem and the application of SLAM to large-scale environments are not addressed. Although it is important to note that SLAM methods that perform well locally can be extended to large-scale problems using different approaches, such as global mapping [19] or global optimization [20].

On the other side, when an attitude measurement $y^N{}_a$ is available, the system state is updated. Since most low-cost AHRS devices provide their output in Euler angles format, the following measurement prediction model $h_a = h(\hat{x}_v)$ is used:

$$\begin{bmatrix} \theta_v \\ \phi_v \\ \psi_v \end{bmatrix} = \begin{bmatrix} \mathrm{a\ tan\ 2}(2(q_3q_4 - q_1q_2), 1 - 2(q_2^2 - q_3^2)) \\ \mathrm{a\ sin}\ (-2(q_1q_3 - q_2q_4)) \\ \mathrm{a\ tan\ 2}(2(q_3q_4 - q_1q_4), 1 - 2(q_3^2 - q_4^2)) \end{bmatrix} \tag{9}$$

During the initialization period, position measurements y_r are incorporated into the system using the simple measurement model $h_r = h(\hat{x}_v)$:

$$h_r = \left[p_x, p_y, p_z\right]^T \tag{10}$$

The regular Kalman update equations [7, 16, 18] are used to update attitude and position whenever is required, but using the corresponding Jacobian ∇H and measurement noise covariance matrix R related to the AHRS observation model.

In this work, the GPS signal is used for incorporating metric information into the system in order to retrieve the metric scale of the estimates. For this reason, it is assumed that the GPS signal is available at least during some period at the beginning of the operation of the system. In order to obtain a proper operation of the system, this period of GPS availability should be enough for allowing the convergence of the depth for some initial features. After this initialization period, the system is capable of operating using only visual information.

In Ref. [21], it is shown that only three landmarks are required for setting the metric scale in estimates. Nevertheless, in practice, there is often a chance that a visual feature is lost during the tracking process. In this case, it is convenient to choose a threshold $n \geq 3$ of features that present convergence in order to end the initialization process. In this work, good experimental results were obtained with $n = 5$.

An approach for testing features convergence is the Kullback distance [22]. Nevertheless, the computational cost of this test is quite high. For this reason, the following criterion is proposed for this purpose:

$$\max(eig(P_{y_i})) < \frac{\|y_i - r^N\|}{100} \tag{11}$$

where P_{yi} is the covariance matrix of the y_i feature, and it is extracted from the system covariance matrix P. If the larger eigenvalue of P_{yi} is smaller than a centime of the distance between the feature and the camera, then it is assumed that the uncertainty of the visual features is small enough to consider it as an initial landmark.

4. Results

In this section, the results related to the proposed visual odometry approach for UAV localization in industrial environments are discussed. The results obtained using synthetic data from simulations as well as that obtained from experiments with real data are presented. The experiments were performed in order to validate the performance, accuracy, and viability of the proposed localization method, considering a real outdoor scenario. Previous works by the authors [7, 18] and other researchers [2] have proven that similar solutions can reach real-time performance, being of more interest development of robust solutions that can be optimized further down the line. The proposed method was implemented in MATLAB©.

http://dx.doi.org/10.5772/intechopen.70005

4.1. Experiments with simulations

The values of the parameters used to simulating the proposed method were taken from the following sources: (i) the parameters of the AHRS were taken from Ref. [8], (ii) the model used for emulating the GPS error behavior was taken from Ref. [23], and (iii) the parameters of the monocular camera are the same for real camera used in real-data experimentation.

In order to validate the performance of the proposal, two different scenarios were simulated (see **Figure 7**) as follows:

1. The environment of the aerial robot is composed of landmarks uniformly distributed over the ground. The quadrotor performs a circular flight trajectory, at a constant altitude, after the taking off.
2. The environment of the aerial robot is composed of landmarks randomly distributed over the ground. The quadrotor performs and eight-like flight trajectory, at a constant altitude, after the taking off.

In simulations, the data association problem is not considered, that is, it is assumed that visual features can be detected and tracked without errors. Also, it is assumed that the aerial robot can be controlled perfectly.

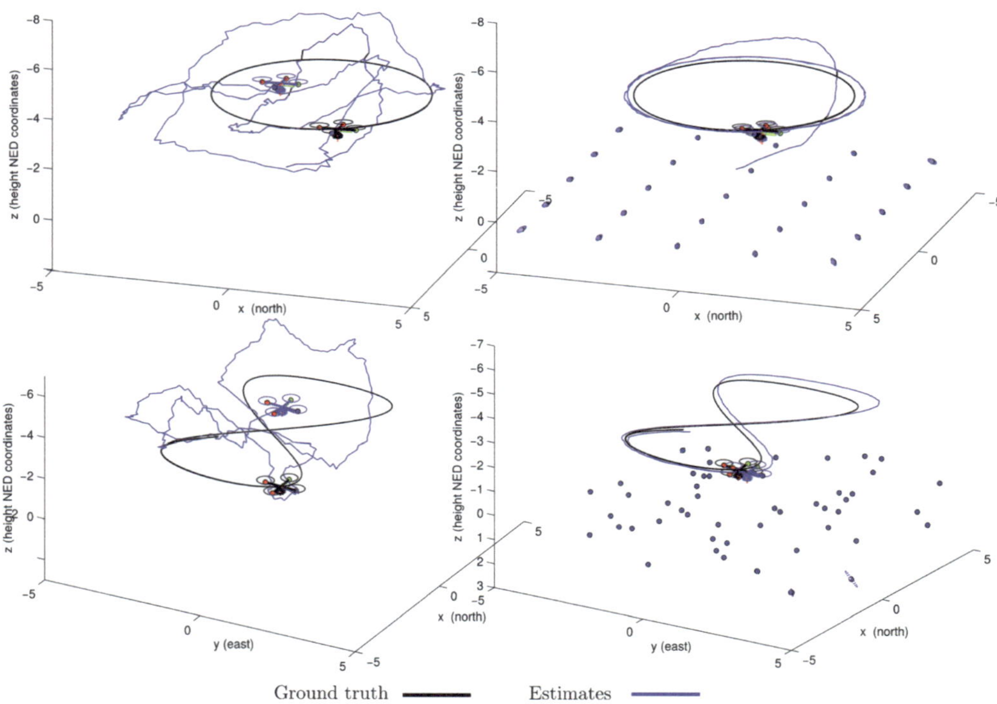

Figure 7. Estimation of two flight trajectories obtained with the filtered GPS data (left plots), and with the scheme proposed of visual-based navigation (right plots).

In order to obtain a statistical interpretation of the simulations results, the MAE (mean absolute error) was computed for 20 Monte Carlo runs. The MAE was calculated for the following cases:

1. Trajectory estimated using only filtered data from the GPS.
2. Trajectory estimated using visual information in combination with GPS data, during all the simulation.
3. Trajectory estimated using visual information and GPS data, but only during the initialization period.

In the results presented in **Figure 8**, it can be appreciated the typical SLAM behavior, every time when the aerial robot flight near to its initial position, when the visual-based scheme is used. In this case, note that when the initial visual features are recognized again, the error is minimized. On the other hand, in the case where GPS data is fused into the system during the whole trajectory, it can be appreciated an influence of the GPS error when the aerial robot flight near to its initial position. In this latter case, the error is less minimized. The above results can suggest that for some scenarios, it is better to navigate relying only on visual information.

Also, it is important to note that for trajectories where the aerial robot moves far away from the initial position, the use of GPS data can be very useful because an upper bound is imposed to the error drift which is inherent to the navigation scheme based only on vision.

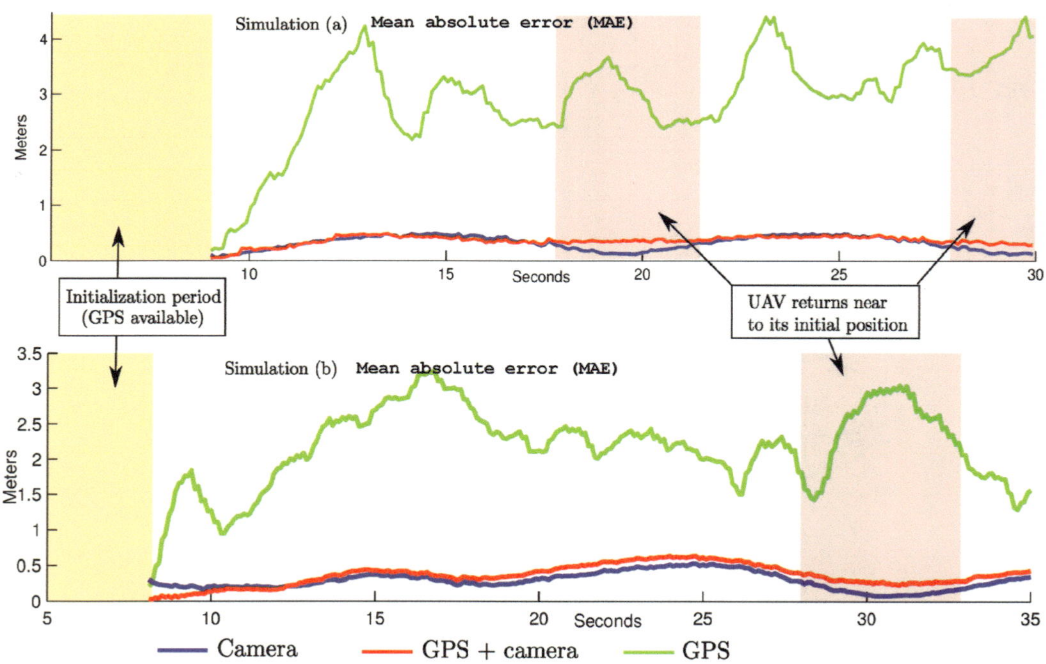

Figure 8. Mean absolute error (MAE) in position computed from two simulations (a and b) out of 20 Monte Carlo runs: (upper plot) simulation (a) results; (lower plot) simulation (b) results.

It is important to note that errors related to the slow-time varying bias of the GPS can be modeled in Eq. (2) by considering a bigger measurement noise covariance matrix. However, in experiments, it was found that if this matrix is increased too much then the convergence of initial visual features can be affected. Future work could include an adaptive approach for fusing GPS data, or for instance, to include the bias of the GPS into the system state in order to be estimated.

4.2. Experiments with real data

The custom build sample collector UAV was used to perform experiments with real data. In experiments, the quadrotor has been manually radio-controlled to capture data. The data captured from the GPS, AHRS, and frames from the camera were synchronized and stored in an ROS data set. The frames with a resolution of 320 × 240 pixels, in gray scale, were captured at 26 fps. The flights of the quadrotor were conducted in industry-like facilities.

The surface of the field is mainly flat and composed of asphalt, grass, and dirt, but the experimental environment also included some small structures and other manmade elements. In average eight to nine GPS, satellites were visible at the same time.

In order to evaluate the estimates obtained with the proposed method, the flight trajectory of the quadrotor was determined using an independent approach. In this case, the position of the camera was computed, at each frame, with respect to a known reference composed of four marks placed on the floor forming a square of known dimensions. The perspective on four-point (P4P) technique described in [24] was used for this purpose.

As it was explained earlier, an initialization period was used for incorporating GPS readings in order to set the metric scale of the estimates. After the initialization period, the estimation of the trajectory of the aerial robot was carried out using only visual information.

The same methodology used with simulated data was employed with real data. Therefore, the same experimental variants were tested: (i) GPS, (ii) GPS + camera, (iii) camera (GPS only during the initialization period). All the results were obtained averaging 10 experimental outcomes. **Figure 9** shows the evolution of the estimated flight trajectory along the time, which was obtained with each experimental variant. **Table 1** shows a summary of the above experimental results. In this case, in order to compute the error in position, the trajectory computed with the P4P technique was used as ground truth.

It worthwhile to note that analyzing the above results, it can be verified similar conclusions that they were obtained with simulations: The use of GPS exclusively can be unreliable for determining the vehicle position in the case of fine maneuvers. In flight trajectories near to the initial reference, the error can be slightly low relying only on visual information.

Regarding the application of the proposed method in a real-time context, the number of features that are maintained in the system state is considerably below an upper bound that should allow a real-time performance, for instance by implementing the algorithm in C or C++ languages. In particular, since early works in monocular SLAM as Davison [25], it was

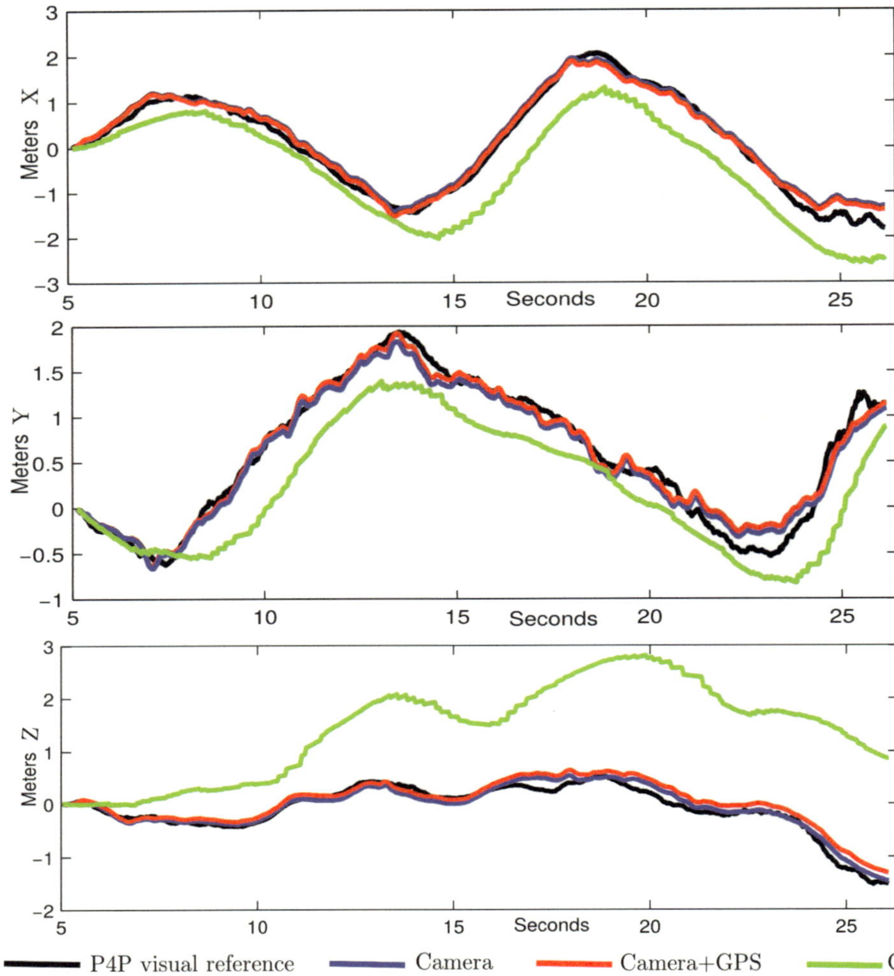

Figure 9. Flight trajectory estimated with: (i) P4P visual reference, (ii) camera, (iii) camera + GPS, and (iv) GPS. The position is expressed in north coordinates (upper plot), east coordinates (middle plot), and altitude coordinates (lower plot). In every case, an initialization period of 5 s was considered with GPS availability.

Method	NOF	aMAE
GPS	–	$1.70 \pm 0.77\sigma$
Camera + GPS	$56.4 \pm 10.2\sigma$	$0.21 \pm 0.11\sigma$
Camera	$57.9 \pm 9.3\sigma$	$0.20 \pm 0.09\sigma$

NOF stands for the average number of visual features contained within the stochastic map. The aMAE is the average mean absolute error in position (in meters).

Table 1. Summary of results with real data.

shown the feasibility of real-time operation for EKF-based methods for maps composed of up to 100 features.

5. Conclusions

An industrial system to automatize the water sampling processes on outdoor basins in a wastewater treatment plant has been proposed, with novel research to solve the localization for navigation problem under unreliable GPS. The architecture of the whole system has been described, while specifications at the hardware level have been presented for those elements designed completely, including the sample collector UAV, and the proposed localization technique has been described and validate with experiments performed over simulated and real data. The localization technique presented can be described as a vision-based navigation and mapping system applied to UAV.

The proposed estimation scheme is similar to a pure monocular SLAM system, where a single camera is used for estimating concurrently both the position of the camera and a map of visual features. In this case, a monocular camera equipped in the aerial robot, which is pointing to the ground, is used as the main sensory source. On the other hand, in the proposed scheme additional sensors, which are commonly available in this kind of robotic platforms, are used for solving the typical technical difficulties which are related to purely monocular systems.

One of the most important challenges, regarding the use of monocular vision in SLAM, is the difficulty in estimating the depth information of visual features. In this case, a method based on a stochastic technique of triangulation is proposed for this purpose.

Perhaps the other most relevant challenge, regarding the use of monocular vision, has to do with the difficulty of retrieving the metric scale of estimates. In this work, it is assumed that the GPS signal is available during an initial period of time which is used for set the metric scale of the estimates. After the initialization period, the proposed system is able to estimate the position of the flying vehicle using only visual information.

For some scenarios, it was seen that the exclusive used of filtered GPS data can be unreliable for performing fine maneuvers. This is due to the noise implicit in GPS measurements. In this context, the following conclusions were found based on the experiments with real data as well as with simulations:

1. Even the use of very noisy GPS data, during an initial short period of time, can be enough for set the metric scale of the estimates obtained by a monocular SLAM system.

2. The integration of GPS measurements can be avoided for flight trajectories near to the origin of the navigation frame.

Acknowledgements

This research has been funded by EU Project AEROARMS project reference H2020-ICT-2014-1-644271, http://www.aeroarms-project.eu/.

Author details

Rodrigo Munguia[1], Edmundo Guerra[2], Sarquis Urzua[1], Yolanda Bolea[2] and Antoni Grau[2]*

*Address all correspondence to: antoni.grau@upc.edu

1 Department of Computer Science, CUCEI, University of Guadalajara, Guadalajara, Mexico

2 Automatic Control Department, Technical University of Catalonia, Barcelona, Spain

References

[1] Ping JTK, Ling AE, Quan TJ, Dat CY. Generic unmanned aerial vehicle (UAV) for civilian application—A feasibility assessment and market survey on civilian application for aerial imaging. In: 2012 IEEE Conference on Sustainable Utilization and Development in Engineering and Technology (STUDENT). 2012. pp. 289-294

[2] Davison AJ, Reid ID, Molton ND, Stasse O. MonoSLAM: Real-time single camera SLAM. IEEE Transactions on Pattern Analysis and Machine Intelligence. 2007 Jun;29(6):1052-1067

[3] Ganganath N, Cheng CT, Tse CK. A constraint-aware heuristic path planner for finding energy-efficient paths on uneven terrains. IEEE Transactions on Industrial Informatics. 2015 Jun;11(3):601-611

[4] Meier L, Honegger D, Pollefeys M. PX4: A node-based multithreaded open source robotics framework for deeply embedded platforms. In: 2015 IEEE International Conference on Robotics and Automation (ICRA). 2015. pp. 6235-6240

[5] Quigley M, Conley K, Gerkey B, Faust J, Foote T, Leibs J, et al. ROS: An open-source robot operating system. In: ICRA Workshop on Open Source Software. 2009. p. 5

[6] MAVLink Micro Air Vehicle Communication Protocol—QGroundControl GCS [Internet]. Available from: http://qgroundcontrol.org/mavlink/start [Accessed: January 26, 2017]

[7] Munguía R, Grau A. Monocular SLAM for visual odometry: A full approach to the delayed inverse-depth feature initialization method. Mathematical Problems in Engineering. pp. 1-26; 2012

[8] Munguía R, Grau A. A practical method for implementing an attitude and heading reference system. International Journal of Advanced Robotics Systems. 2014 Apr 29;11(4):62

[9] Wang M, Yang Y, Hatch RR, Zhang Y. Adaptive filter for a miniature MEMS based attitude and heading reference system. In: PLANS 2004 Position Location and Navigation Symposium (IEEE Cat No04CH37556). 2004. pp. 193-200

[10] Jurman D, Jankovec M, Kamnik R, Topič M. Calibration and data fusion solution for the miniature attitude and heading reference system. Sensors and Actuators A: Physical. 2007 Aug 26;138(2):411-420

[11] Spilker Jr. JJ, Axelrad P, Parkinson BW, Enge P. Global Positioning System: Theory and Applications, Volume I [Internet]. American Institute of Aeronautics and Astronautics; 1996. 1. (Progress in Astronautics and Aeronautics). Available from: http://arc-test.aiaa.org/doi/book/10.2514/4.866388 [Accessed: January 27, 2017]

[12] Grewal MS, Weill LR, Andrews AP. Global positioning systems, inertial navigation, and integration. New York: John Wiley; 2001. p. 392

[13] GPS: Essentials of Satellite Navigation: Compendium : Theories and Principles of Satellite Navigation, Overview of GPS/GNSS Systems and Applications. U-Blox; 2009. p. 176

[14] Shi J, Tomasi C. Good Features to Track. 1994. pp. 593-600

[15] Davison AJ, Murray DW. Mobile robot localisation using active vision [Internet]. Springer; 1998. Available from: http://link.springer.com/chapter/10.1007/BFb0054781 [Accessed: September 30, 2015]

[16] Munguia R, Grau A. Monocular SLAM for visual odometry. In: IEEE International Symposium on Intelligent Signal Processing, 2007 WISP 2007. 2007. pp. 1-6

[17] Munguía R, Grau A. Concurrent Initialization for bearing-only SLAM. Sensors. 2010 Mar 1;10(3):1511-1534

[18] Guerra E, Munguia R, Bolea Y, Grau A. New validation algorithm for data association in SLAM. ISA Transactions. 2013 May 20;52:662-671

[19] Munguia R, Grau A. Closing loops with a virtual sensor based on monocular SLAM. IEEE Transactions on Instrumentation and Measurement. 2009;58(8):2377-2384

[20] Strasdat H, Montiel JMM, Davison AJ. Real-time monocular SLAM: Why filter? In: 2010 IEEE International Conference on Robotics and Automation (ICRA). 2010. pp. 2657-2664

[21] Belo FAW, Salaris P, Fontanelli D, Bicchi A. A complete observability analysis of the planar bearing localization and mapping for visual servoing with known camera velocities. International Journal of Advanced Robotic Systems. 2013 Apr 17;10(4):197

[22] Bailey T. Constrained initialisation for bearing-only SLAM. In: 2003 Proceedings ICRA'03 IEEE International Conference on Robotics and Automation [Internet]. 2003. pp. 1966-1971. Available from: http://ieeexplore.ieee.org/xpls/abs_all.jsp?arnumber=1241882 [Accessed: June 4, 2013]

[23] Rankin J. An error model for sensor simulation GPS and differential GPS. In: IEEE Position Location and Navigation Symposium, 1994. 1994. pp. 260-266

[24] Chatterjee C, Roychowdhury VP. Algorithms for coplanar camera calibration. Machine Vision and Applications. 2000;12(2):84-97

[25] Davison AJ. Real-time simultaneous localisation and mapping with a single camera. In: IEEE International Conference on Computer Vision. 2003. pp. 1403-1410

Section 3

Applications

Chapter 7

On-Board High-Performance Computing For Multi-Robot Aerial Systems

Leonardo Camargo Forero, Pablo Royo and Xavier Prats

Additional information is available at the end of the chapter

http://dx.doi.org/10.5772/intechopen.69443

Abstract

With advancements in low-energy-consumption multi/many core embedded-computing devices, a logical transition for robotic systems is *Supercomputing*, formally known as high performance computing (HPC), a tool currently used for solving the most complex problems for humankind such as the origin of the universe, the finding of deceases' cures, etc. As such, HPC has always been focused on scientific inquires. However, its scope can be widening up to include missions carried out with robots. Since a robot could be embedded with computing devices, a set of robots could be set as a cluster of computers, the most reliable HPC infrastructure. The advantages of setting up such an infrastructure are many, from speeding up on-board computation up to providing a multi-robot system with robustness, scalability, user transparency, etc., all key features in supercomputing. This chapter presents a middleware technology for the enabling of high performance computing in multi-robot systems, in particular for aerial robots. The technology can be used for the automatic deployment of cluster computing in multi-robot systems, the utilization of standard HPC technologies, and the development of HPC applications in multiple fields such as precision agriculture, military, civilian, search and rescue, etc.

Keywords: high performance computing (HPC), cluster of computers, multi-robot systems, HPC-ROS, ubiquitous supercomputing, unmanned aerial system (UAS), unmanned aerial vehicle (UAV), HPC cluster of robots

1. Introduction

High performance computing (HPC) can be thought as a tool for allowing or speeding up the solving of problems that either cannot be solved in a regular desktop computer or it would take too much time to do it. Within the scope of robotics, for example in missions carried

INTECH
open science | open minds

© 2017 The Author(s). Licensee InTech. This chapter is distributed under the terms of the Creative Commons Attribution License (http://creativecommons.org/licenses/by/3.0), which permits unrestricted use, distribution, and reproduction in any medium, provided the original work is properly cited.

with unmanned aerial systems (UASs), HPC approaches and technologies could be used for distributed on-board computing, in order to maximize efficiency.

Among many existing HPC infrastructures, a cluster of computers is perhaps the most common. The idea consists on integrating multiple computing units (e.g., CPUs, GPUs, desktop computers, servers, etc.) into a single platform where software can use the entire resources acting as one, therefore, facilitating a timely execution. To do so, the software must be written with specific parallel or distributed software libraries in order to exploit the resources among which it is being executed. For example, parallel software on-board robots could exploit embedded CPU cards with multiple cores, etc., even among independent robots acting as a cluster of computers.

To sustain the objective of speeding up computation, given the fact that a cluster of computers is build up out of multiple independent computing units, certain features must be provided by the platform itself. Such features are scalability, failure tolerance, robustness, reliability, and user transparency. For the rest of this document, these features are referred as HPC features.

Scalability refers to the idea of adding or subtracting computing units to the cluster (nodes), without compromising the entire system. In the context of robotic systems performing real-world missions, scalability can be used to adapt the mission objectives to the available robots, providing the capability of adding or subtracting new robots if needed.

Failure tolerance refers to the capacity of recovering in case of failures such as those occurring in nodes, software, etc. Given the changing environment conditions, in which real-world robotic missions are carried out, different sources of failures are expected. Tolerance to such failures, as in load redistribution, for example, is a much desired feature for robotic systems.

Robustness refers to the capacity of adaptation to possible errors that might occur during software execution, where the probability of such occurrence is higher, in comparison with sequential (in a single computing unit) execution.

Reliability relates to the results obtained by a software execution. This is, software output, obtained through distributed/parallel approaches must be the same as it would be in a sequential approach. Both robustness and reliability are much desired for software embedded on robotic systems.

User transparency refers to the fact that the user interacting with the HPC infrastructure does not need to understand the complexity necessary to set a single computing facility built up of distributed independent computing units. This offers a whole set of options to robotic systems. For example, a parallel application could be developed to pilot all robots at once.

Furthermore, time efficiency and task distribution can also be considered as HPC features.

The HPC features are provided by the platform itself and by parallel software libraries such as message passing interface (MPI) [1]. The platform is depicted as a set of specific software layers with particular functions and multiple open-source and privative software solutions (for each layer)–HPC software layers. These layers are operating system, user system/file system, batch system, and applications.

Figure 1 depicts the organization of the HPC software layers in a traditional cluster of computers (a) and the proposal for replication of such layers in a cluster of computers built with robots (b), in this case UASs, as presented in the following sections.

As it can be seen in **Figure 1**, nodes in an HPC cluster can be classified as master and slaves. The master node represents the machine in charge of managing the entire infrastructure, monitor it, control it, and serving as an access point for the users. The slaves nodes are the ones in charge of performing computation.

Theoretically, any type of computing unit could potentially be part of an HPC cluster of computers, provided that such software layers can be deployed over it. Such idea extends to robots.

While, nowadays, the computing capability of a common HPC node far exceeds those of a robot, the idea of splitting a software task and performing a parallel/distributed approach holds. Moreover, the tasks being performed by a robot are not as complex as it would be a task performed by a common HPC node.

And since an operating system can be installed in computing cards that can be implemented on robotic infrastructures, theoretically, an HPC cluster of computers could be deployed with multiple robots.

Attempts to do so have been done in the past. Holland et al. [2] proposed the UltraSwarm system, with the objective of providing an unmanned aerial vehicle (UAV) swarm with a single controlling intelligence based on HPC.

The UltraSwarm system is a combination of swarm intelligence and wireless cluster computing based on Linux and Bluetooth communications. However, the authors did not implement the HPC cluster computing software layers, which ultimately leads to not being able of offering user transparency, scalability, failure tolerance, robustness and reliability, in the sense of classical cluster computing.

Marjovi et al. [3] proposed the concept of robotic cluster and defined as: A robotic cluster is a group of individual robots, which are able to share their processing resources among the group in order to quickly solve computationally hard problems.

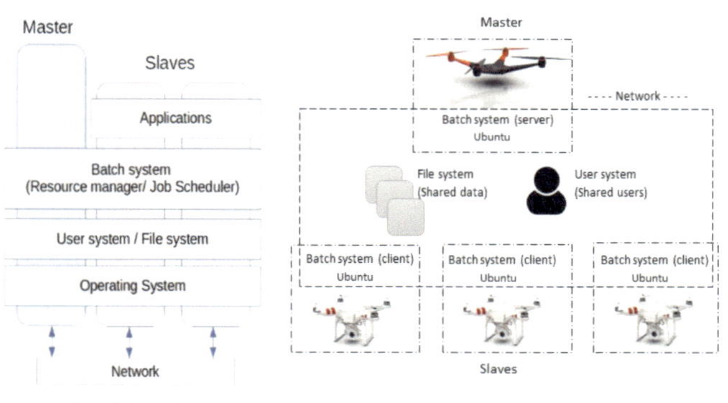

Figure 1. HPC cluster of computers software layers.

The authors implemented a software platform that resembles the one here presented, specifically to what they called the middleware layer. In their case, the HPC libraries MPI, which allows software to run in parallel and offers certain features to handle communication among distributed processes.

As for the concept of middleware, as a software layer, between two software layers (for this case: the operating system and the parallel applications layer) and with the objective of seeing the cluster's computing units as a single one, it is valid to consider MPI as such. However, it is perhaps more adequate to consider the batch system (**Figure 1**) as the HPC cluster middleware and allocate MPI libraries to the applications layer, as it is, in the traditional concept of an HPC cluster of computers. Moreover, the authors Marjovi et al. [3] did not implement the user system/file system layer, which is the one providing an HPC cluster with the ability of replicating data among the independent computing units and to set a multipurpose/multiuser infrastructure.

Moreover, a single robot (e.g., an UAS) could be considered as an HPC infrastructure in itself. Multiple and different types of computing boards (e.g., CPUs, GPUs, FPGAs) could be embedded on it, giving the robot the ability of locally processing data and even doing it using parallelism (e.g., open MP [4], open MPI [5] or compute unified device architecture–CUDA [6]). This could reduce the dependency on data links, favoring therefore mission real-time decisions.

Salamí et al. [7] implemented two applications: a parallel hot spot detection algorithm and a jellyfish detection algorithm on board unmanned aerial system (UAS) for fast response using and comparing three different computing cards (EPIA N700-15 VIA C7-1.5, Pandaboard with OMAP 4430 integrated CPU, INTEL T7500 on GENE-9655 motherboard and TilenCore-Gx board with TILE-Gx36). The results showed that performing on-board parallel computing provides faster response time for critical missions.

Multiple companies provide nowadays multi-core light-weight low-energy-consumption computing cards for embedded robotic systems (e.g., Intel, NVIDIA, Texas Instruments, Raspberry PI Foundation, Qualcomm, Parallela, etc.), ranging from dual core CPU cards like the Intel Edison architecture [8] up to the 256 NVIDIA CUDA cores Jetson TX1 [9].

With advancements in energy consumption decreasing, GPUs represent an important technology for robotic-embedded parallel computing. Jeon et al. [10] presented a real-time vision-guided control algorithm (detection and tracking) for an UAV using CUDA-enabled GPU cards. Cocchioni et al. [11] also showed the implementation of visual-based landing for an unmanned quadrotor using a NVIDIA Tegra Jetson K1.

Furthermore, a cluster of computers can be set up in a single robot as well, as Ribeiro et al. [12] proposed by designing a robotic flying crane with an embedded cluster of computers composed of ODROID-x2 processors [13], Linux Ubuntu and sun grid engine (SGE) [14] as batch system.

Considering multiple robots, an HPC cluster of computers could be set up among them. However, most studies on multi-robot systems are based on swarming approaches focusing mainly on collision avoidance, motion coordination, flying formation, etc., all applications that could be allocated to the applications layer of the HPC cluster of computers platform.

Multiple research has been done using robot operating system (ROS) [15] for robot swarming [16–22] but not setting up an HPC cluster of computers (HPC software layers). It is our belief that by setting up a cluster of computers with a multi-robot (e.g., UAS) system (*HPC cluster of robots*), several advantages appear in comparison with using a single UAS. The following lists such advantages, relating them with the HPC features (S = scalability, FT = failure tolerance, RO = robustness, RE = reliability, UT = user transparency, TE = time efficiency, TD = task distribution) and the HPC software layers:

1. Cooperative/distributed on-board mission execution (TE, TD) (e.g., mission on-board software). Some mission examples are tracking and target recognition, vegetation health analysis, hot-spot detection and extinction, surveillance and intelligence, etc. This is managed by the applications and the batch system layers.
2. Distribution of mission area to multiple UASs (i.e., each UAS is assigned with a mission subarea) (TE, TD). This is managed by the applications and the batch system layers.
3. Single access point (UT). The master node, in a cluster of computers, acts as user frontend. In this sense, once a user is logged in, the master can access the entire infrastructure, as it would be a single computing unit made up of multiple computing units. For example, a parallel multi-UAS autopilot could be developed or even a pilot could control the entire system by controlling only the master.
4. Multipurpose/multiuser systems. A cluster of computers is designed to allow multiple users execute multiple purposes software and doing it so efficiently. This is one of the specific objectives of the batch system and its interaction with the user and file system.
5. Lower system implementation costs (S, FT). Deploying a single UAS with high computing power will require higher costs that several UASs with little computing power acting as a cluster of computers.
6. No single point of failure (FT). In case of a single UAS, a failure could compromise the entire system.
7. Node addition in case of need (S). This is managed by the batch system.
8. Workload reassignment in case of software failure, node failure, or even robot destruction (intentionally or unintentionally) (S, FT, RO). This is managed by the batch system.

High performance computing has been used, traditionally, as a tool for solving the most complex problems known to humankind. However, recently, the emergence of the ideas behind big data expands HPC scope from sciences to commercial applications. The infrastructure underneath big data software is in fact a supercomputer with similar layers as the HPC software layers. The same applies for cloud computing, grid computing, and opportunistic computing. In this sense, supercomputing is everywhere, except in robots yet.

With regard to the proposed definition of robotic cluster [3], it is clear that the driver of implementing HPC infrastructures remains as the capacity of quickly processing computationally hard problems (computing efficiency), according to the authors. In fact, it is fair to say that HPC has always been targeted as a tool for such purpose.

However, in this work, it is proposed to put such main driver aside and focus on the HPC features used to provide it. By exploiting such features, the potential of developing and creating complex applications is enormous. Applications running over multiple independent computing units, acting as one (an abstraction of user transparency) and enjoying high computing efficiency.

To do so, it is proposed the development of an HPC enabler, a platform middleware technology to bring HPC features everywhere—*Ubiquitous supercomputing*. Not only to common HPC cluster nodes or robots but also to any device capable of supporting it.

Such enabler should allow (HPC enabler features):

1. The automatic deployment (installation and configuration) of the HPC software layers.
2. The use of standard HPC software applications for each layer.
3. Serve as a middleware where users and researchers can develop software applications to be deployed on the HPC applications layer (uppermost) relying on scalable subjacent software layers, and;
4. Adaptation to the specific conditions of the devices and the missions carried out by such devices. For example, the nodes in an HPC cluster of computers are normally connected through high-speed high-bandwidth low-latency communication technologies. Such conditions are not available for robotic applications nowadays.

In this work, it is presented as the high-performance computing/robot operating system (HPC-ROS) package, an enabler technology for ubiquitous supercomputing.

Robot operating system (ROS) is a middleware to facilitate robotics systems development, easing technology reutilization, and research cooperation via packages. As far as we can tell, there is neither a single available package in the current ROS version (KINETIC) nor the previous versions package repositories, that provides the features proposed by the HPC-ROS package presented here.

Following, Section 2 introduces a novel conceptual description to implement HPC robotic applications in the future. In Section 3, the HPC-ROS package is presented and Section 4 depicts current conclusions.

2. Conceptual description

In this section, it introduced a conceptual description (ontology) to facilitate the design of missions carried out with HPC cluster of robots.

Considering the idea of a robot as a software container, a mission can be understood as a set of software executed in the robots and in ground stations. This software could be of two types: *control* and *mission*. Control refers to the software used for the management of the robot, and it is classified in three types, as:

- Operation: e.g., flying (telemetry, autopilot, etc.).
- Payload: Sensor, actuators management software.
- Awareness: Software for the interaction with the environment and other actors e.g., Air traffic management (ATM).

Mission refers to the embedded software used for the carrying of the mission.

Under these ideas, the concept of *HPC job* is presented. An HPC job is a set of software that requires a set of hardware resources for its execution (e.g., quantity of CPUs, RAM, GPUs, disk space, etc.) during a time window known as walltime.

The conceptual description, here presented, serves two purposes:

1. Facilitate the design of scalable missions carried out by HPC cluster of robots
2. Mapping of a mission into HPC jobs.

The following terms are introduced within the scope of this conceptual description.

A *system* is a set of independent heterogeneous or homogeneous nodes (e.g., robots, ground stations, computers, etc.) performing a set of missions distributively and/or cooperatively.

A *mission* is a set of HPC jobs running on an infrastructure.

An *infrastructure* is the system's subset of nodes assigned to a mission.

The following features are part of this conceptual description:

1. A system is composed of 1 up to N missions.
2. A mission is composed of 1 up to M Jobs (HPC Jobs) and an infrastructure.
3. A job is composed of 1 up to X software applications.
4. Software is classified as payload or operation.
5. Each software has a required amount of resources (e.g., CPU cores, GPU cores, RAM, etc.) and a walltime (estimated time of completion).
6. An infrastructure is an HPC cluster, and therefore, it implements all the HPC layers and at least a master node and a set of slave nodes.
7. Each system, mission, job, software, infrastructure, cluster, node has a cost, that can be used for inter (multi-system) or intra (mission-wise, cluster-wise, job-wise, etc.) system negotiation.

Moreover, the nodes within a system are classified, according to the *robot role description (RRD)* concept introduced here, as:

1. Master: This is the HPC cluster (infrastructure) master, the machine in charge of implementing the server-side of each HPC software layer. The master node acts as the frontend, to which users interact via HPC jobs submitted to the server-side of the batch system. Normally, though not necessarily, the master node does not perform computation, and therefore, it does not require implementing powerful computing resources such as CPUs, etc.

2. Shadow master: In case of master failure, a shadow master configured as AP (Active/Passive) or AA (Active/Active–load balancing) is set in an HPC cluster.

3. Slave: The slave nodes are the ones in charge of performing computation. The client-side of each HPC software layer is set upon slave nodes.

4. Cluster operator: Considering the autonomy level of a robot, some nodes might require a human operator, pilot, etc. However, relying on the HPC software layers, software in the applications layer could be deployed to operate the entire cluster. This node is the pilot/cluster interface. This is, the pilot interacts with the cluster operator, which in turn commands the entire cluster. A complete scheme can be devised using the file system to handle multiple user profiles and node operators.

5. Mission controller: This is the node, which controls the mission. It is the node, where a software application, in the applications layer, is deployed to do so. In this sense, this node belongs to the infrastructure assigned to the mission. This is a node whose scope is intra-cluster.

6. System controller: This is the node, which controls the system. It is the node, where a software application, in the applications layer, is deployed to do so. In this sense, this node could belong to the infrastructure of a particular mission or be an independent node. This is a node whose scope is inter-cluster.

7. User-defined roles: To support all type of missions, a node can be assigned a user-defined role e.g., fire-detector, fire-extinguisher, etc.

The RRD allows for the combination of multiple roles in a single node. For example, the master node of mission infrastructure could act as mission or system controller. However, a system controller and a mission controller must exist in every system no matter the amount of nodes, missions, etc.

Moreover, an HPC cluster of computers can be set up in a single node, considering that multiple computing cards or even complete computers could be embedded in a robot. This way, using multiple embedded computers, multiple roles could be assigned to a single node.

The batch system is a very configurable software layer that can be used to set up and manage computing or noncomputing resources, with different computing queues (e.g., queue for nodes with GPUs, queue for nodes with FPGAs, etc.), priorities (per user-profile, per-group, per-user, per-job-type, etc.), simple (e.g., FIFO), or complex prioritization schemes such as allowing a second-priority job to run first if the required resources for the first-priority job are not available and only during the time such resources are busy. By combining the RRD and the batch system and subjacent layers, it is possible to devise very complex and efficient systems, encouraging all type of imaginable missions.

To clarify these ideas, it is introduced the following example. Consider a system called *Mars rover/UAS cluster of robots*.

The *system* is a set of two UASs and two Mars Rovers, whose objective is to perform *in-situ* analysis of Martian ground. Both UASs are not necessarily homogeneous in regards of hardware (computing cards, sensors, actuators, etc.) or software (in the application layers). Same applies for the two rovers. However, all four robots have Ubuntu as operating system installed in their computing cards. The system is composed of two missions:

1. *Air*: This mission consists of one HPC job; aid the rovers to move efficiently and safely. To do so, on-board software for 3D mapping is executed. The *infrastructure* for this mission is the set of UASs.
2. *Ground*: This mission consists of HPC parallel jobs for analysis of Martian ground samples. The *infrastructure* for this mission is the set of rovers.

Following *RRD* for this system is presented. This is one of many possible configurations.

RRD per mission is

1. *Air mission*: There is an *UAS master* and two *UAS slaves*. The master acts as slave. There is no *cluster operator*. A human operator controls each UAS. The slave acts as *shadow master* in AP mode.
2. *Ground mission*: There is a *rover master* and two *rover slaves*. The master acts as slave. There is no *cluster operator*. A human operator controls each rover. The slave acts as *shadow master* in AP mode.

RRD per system is:

1. *System controller*: Earth ground station. A node is not necessarily a robot. These ideas are extensible to all sorts of devices.
2. *Mission controller* (Air and Ground): Mission masters.

Some features of this system are:

1. The system controller (Earth ground station) interacts with the missions' controllers
2. The missions' controllers (master UAS and master rover) interact with each other to share the results of the cluster's HPC jobs. The UAS's cluster calculates the location to which the rover's cluster must move. The UAS master shares this information with the rover master, which shares it with the slave rover.
3. The rovers, once knowing the location to which they should move, are directed there, take the ground samples and process them with its computing cards using parallel algorithms (executed on both rovers at the same time).

All the previous information relates to the first purpose of this conceptual description: Facilitate the design of scalable missions carried out by HPC cluster of robots. Regarding the second purpose: Mapping of a mission into HPC jobs, **Figure 2** depicts such mapping.

According to the ontology here presented, a mission is described as a workflow composed of HPC jobs, where each HPC job is a workflow of software applications, executed into the mission infrastructure (cluster of robots).

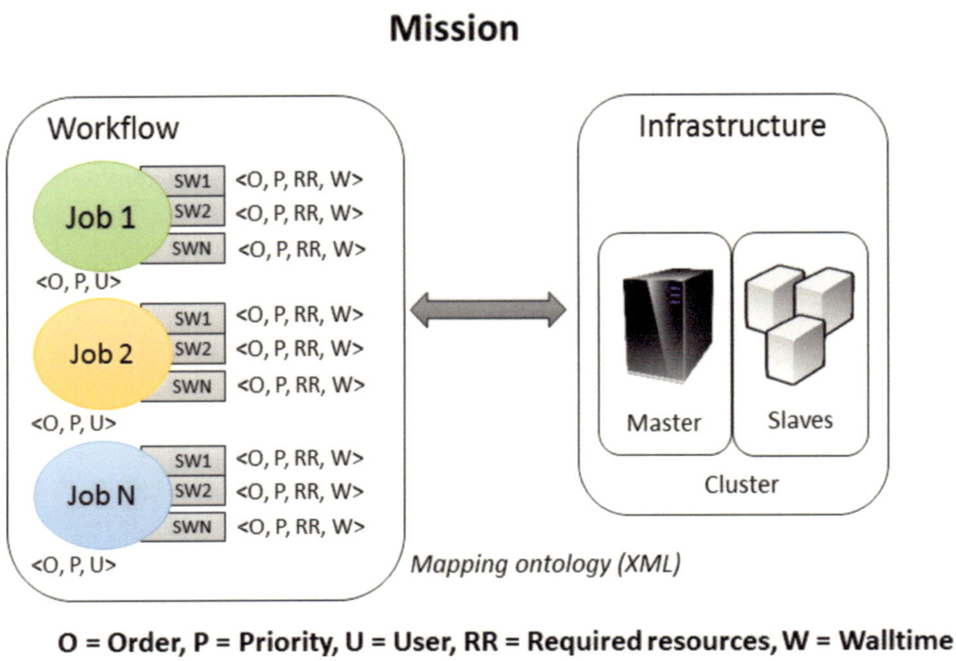

Figure 2. Mission workflow to HPC job mapping.

Each job is executed by a user (U), it has a priority (P), and an order (O). The priority is used by the batch system to decide which job to execute first in the infrastructure. The order is used to describe parallel jobs, jobs that must be executed at the same time. In the case of software applications (SW), order and priority also applies, plus each one has a set of required resources (RR) and a walltime (W). This information is introduced in an XML file.

Following Section 3 introduces the *HPC-ROS package*, an enabler technology for ubiquitous supercomputing.

3. Design and implementation details

This section introduces the *HPC-ROS package*, a technology based on the conceptual description described in Section 2. The HPC-ROS package is a Ubiquitous Supercomputing enabler that can be used to set up HPC cluster of computers in all types of devices (e.g., robots, computers, ground stations, augmented reality devices, etc.) that can be installed with a Linux operating system, Ubuntu for the moment.

The HPC-ROS packages allow (features), as proposed in Section 2:

1. The automatic deployment (installation and configuration) of the HPC software layers: Setting up a cluster of computers specifically the HPC software layers is a complex process that requires deep knowledge of Linux operating systems and HPC in general. Easing its deployment facilitates the robotics community to focus only on the robotics application to be executed on the applications layer of the cluster of computers.

2. The use of standard HPC software applications for each layer: By using standard HPC software applications, normally open-source, an HPC cluster of robots could interact with standard cluster of computers, for example, for missions requiring computing power not available in the robots.

3. Serving as a middleware where users and researchers can develop software applications to be deployed on the HPC applications layer (uppermost) relying on scalable subjacent software layers: Once a HPC cluster of robots is set, researchers could write mission software exploiting parallelism provided by the subjacent layers, and

4. The adaptation to the specific conditions of the devices and the missions are carried out by such devices. For example, the nodes in an HPC cluster of computers are normally connected through high-speed high-bandwidth low-latency communication technologies. Such conditions are not available for robotic applications nowadays: In a traditional cluster of computers, it is imperative that the connectivity between master and slaves is maintained during the execution of a specific software, since all the server side of all HPC software layers is confined to the master node. A common strategy is the deployment of a shadow master. In the case of mobile robots, operating in outdoor environments, a solution to maintain the master-slaves connectivity must be devised. Strategies for such purpose are being currently studied as part of this research work. However, they are not discussed in this chapter.

Moreover, ROS has been widely adopted by the robotics community. It is our belief that developing a ROS package, for the automatic deployment of HPC cluster of robots, is an important contribution that could be used for the development of very complex robotics systems by a community that keeps embracing ROS, more and more as a standard. Nevertheless, the HPC features of the package can be decoupled from ROS easily if the specific robotic system does not use ROS.

The HPC-ROS package, first version (1.0.0) consists of two software components: *HPC deployer component* and *System management component*.

3.1. HPC deployer component

This component relates to features 1, 2, and 3 of the HPC-ROS package. Using this component, the HPC-ROS package installs and configures all software layers in a device installed with Ubuntu. It is currently tested in Ubuntu 14.04, 16.04, MATE, and with ROS Jade version, upon virtual machines and Raspberry PI 3 model B [23].

Figure 3 shows the software architecture of the HPC deployer component. The python scripts (ROS nodes [24]) *Commander* and *Deployer* are used to set different roles in a particular device.

Four options are available for deployment in the current version: *setDefault*, *setMaster*, *setSlave*, and *testCluster*.

- The option setDefault installs and configures base software for high performance computing in Ubuntu, required by all HPC software layers. With this option deployed over a device, the node is capable of performing parallel computing.
- The option setMaster installs and configures the server-side of each HPC software layer configuring therefore a master node.
- The option setSlave installs and configures the client-side of each HPC software layer configuring therefore a slave node.
- Finally, the option testCluster executes a parallel software (MPI) test over the configured cluster.
- All options, except test Cluster, install and configure standard HPC software in each layer.

Figure 4 shows the use of the HPC deployer component for automatic cluster computing setup. Setting up and administrating a cluster of computers is a complex task that requires deep knowledge of Linux OS, networking, information security, parallelization approaches, software architecture, etc. This component facilitates this process so the users or researchers

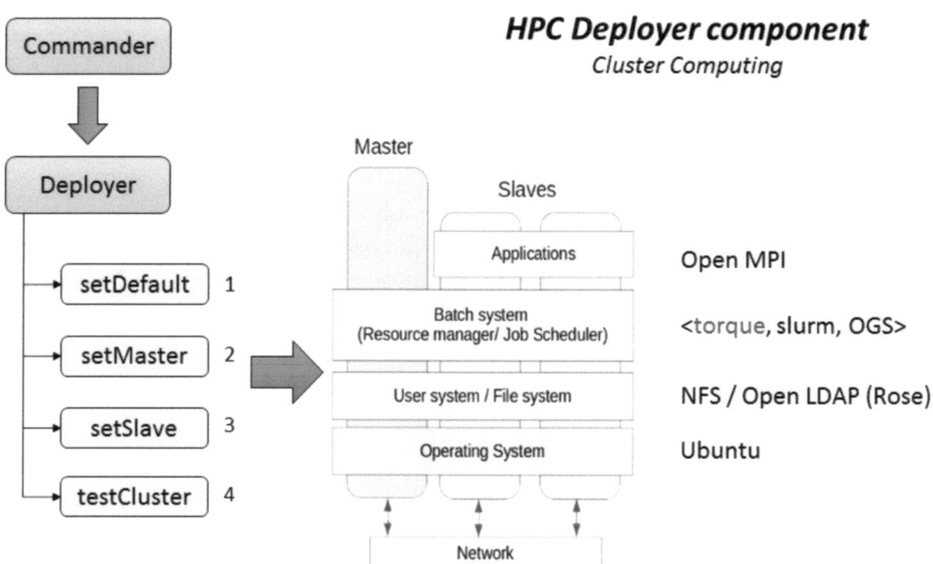

Figure 3. HPC-ROS package–HPC deployer component.

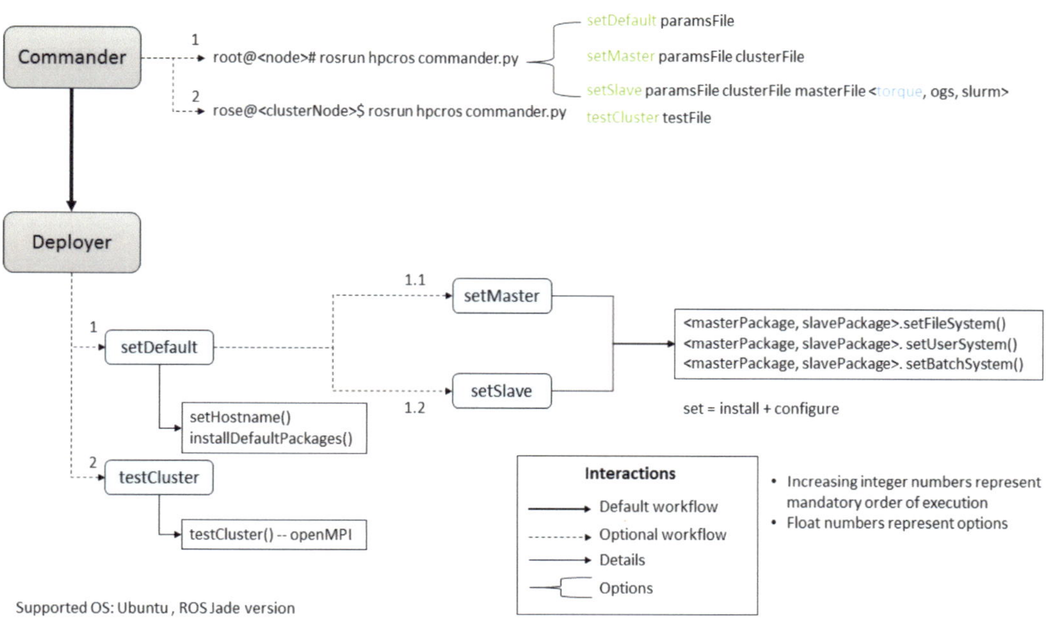

Figure 4. Automatic cluster computing setup.

can focus on the specific software applications (e.g., for robotics) to be implemented upon the applications layer (feature 3 of the HPC-ROS package).

3.2. System management component

This component relates to features 3 and 4 of the HPC-ROS package. This component is currently in design phase. **Figure 5** presents the system management component.

In the current design stage of this component, three subcomponents are foreseen: *dynamic HPC, mission controller*, and *system controller*. The three subcomponents serve two purposes:

1. Act as ROS nodes
2. Provide a set of libraries that can be used by software applications to be deployed upon the application layers

Dynamic HPC—This subcomponent is associated to the HPC features: scalability and failure tolerance. Given the conditions of missions to be carried out by an HPC cluster of robots, where constant connectivity is not granted, four options are available for this subcomponent: *Node addition, Node removal, Cluster connection,* and *Leader selector.*

- The option node addition allows a system to include a node into a specific mission infrastructure. To do so, the dynamic HPC subcomponent interacts with the HPC deployer component.

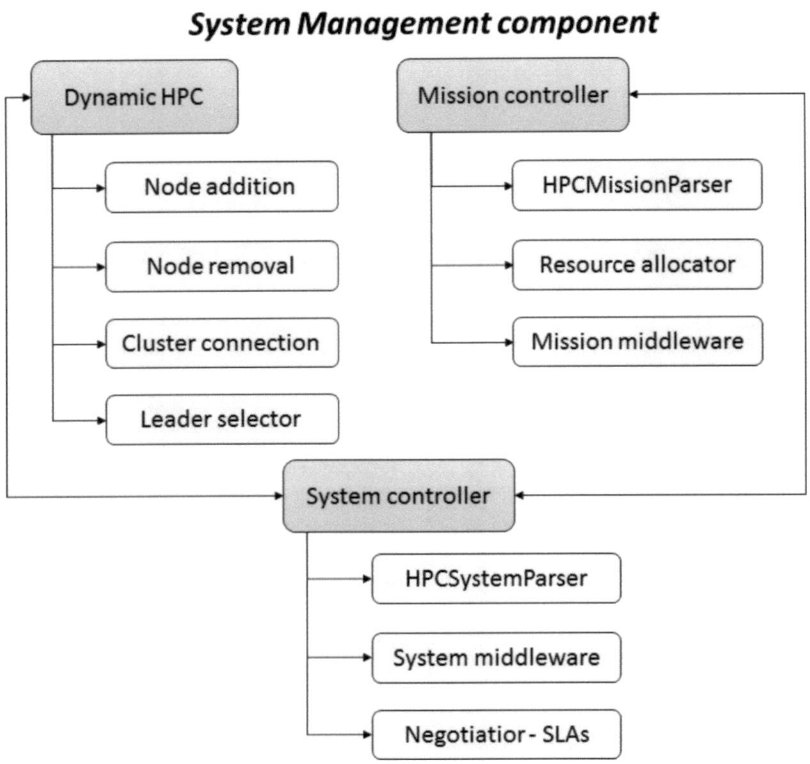

Figure 5. HPC-ROS package–System management component.

- The option node removal allows the removal of a node in case of detected failure, lack of connectivity, etc.
- The option cluster connection facilitates the interaction between HPC software layers in different clusters belonging to the same system or different systems (multi-system level).
- The option leader selector allows selecting a new master node (leader) in case of master and/or shadow master failure.

Mission controller—This subcomponent is to be deployed upon the mission controller node (RRD). It provides three options: *HPCMissionParser*, *Resource Allocator*, and *Mission middleware*.

- The option HPCMissionParser allows mapping of missions into HPC jobs, in order to request transparently (without user-interaction) resources to the batch system. This applies at mission level (intra-cluster)
- The option Resource Allocator facilitates to allocate or reallocate (node adding, node removal) resources to mission jobs.
- The option Mission middleware offers a set of features for message exchange, cooperation, load distribution, etc., between nodes belonging to the same cluster. It relies on the dynamic HPC subcomponent–option cluster connection.

System controller—This subcomponent is to be deployed upon the system controller node (RRD). It provides three options: *HPCSystemParser*, *System middleware*, and *Negotiator–SLAs*.

- The option HPCSystemParser allows mapping of a system into HPC jobs, in order to request transparently (without user-interaction) resources to the batch system of each infrastructure within the system. This applies at system or multi-system level.
- The option System middleware offers a set of features for message exchange, cooperation, load distribution, etc., between clusters belonging to the same system or different systems. It relies on the dynamic HPC subcomponent–option cluster connection.
- The option Negotiator–SLAs allows to negotiate adding of nodes to specific infrastructures, using an infrastructure for the execution of particular jobs, use of resources belonging to a node, a cluster, etc.

This negotiation is based on a cost function, which considers aspects such as: computing capacity (e.g., floating operations per second–FLOPS, quantity of cores, etc.), max endurance (e.g., for the UASs), max range, weight, energy consumption, price, etc.

Each system, mission, job, software, infrastructure, cluster, or node has a cost that can be used for inter (multi-system) or intra (mission-wise, cluster-wise, job-wise, etc.) negotiation.

Using this option, a whole new range of missions could emerge from the use of robotic systems belonging to different companies, organizations, etc. by the implantation of service level agreements (SLAs).

4. Conclusions

The conceptual description presented in Section 2 and the HPC-ROS package in Section 3 could be used to build very complex systems, carrying out missions with multiple heterogeneous components such as UASs, UGSs (rovers), Internet of things devices, sensors, etc., acting as a single distributed collective entity, as a cluster of computers is (collection of multiple heterogeneous computing units).

Two aspects are being addressed with the development of the HPC-ROS package. Quickly and easily deploying a HPC cluster of robots and maintaining its virtual existence (master-slaves connectivity) during the execution of a mission.

Deploying a cluster of computers is not an impossible task but it requires deep knowledge of Linux operating systems, understanding of the ideas behind HPC and ultimately time effort. In this sense, the HPC-ROS package *HPC deployer component* facilitates this task by automating the deployment (installation and default configuration) of a cluster of computers. It can do so in any computing cards installed with Ubuntu.

Current communications technologies prevent a fully operating standard cluster of computers (high bandwidth, low latency, constant connectivity, etc.) to be deployed with multiple robots operating outdoors (e.g., outside a Wi-Fi area). Locally (e.g., in one UAS), cluster computing could be easily implemented (e.g., multiple computing cards embedded on an UAS).

However, the HPC-ROS package aims at multi-robot systems, and it is our belief that in the following years, communication technologies will be ready for mobile embedded HPC. Nevertheless, the HPC-ROS package *System management component* looks for adapting to nowadays communication technologies. We are exploring strategies to maintain the master-slaves connection, at all time, during the execution of a mission like bioinspired models facilitating, therefore, HPC cluster of robots operating outdoors.

Acknowledgements

The authors would like to thank the Colombian government, in particular to the Colombian Administrative Department of Science, Technology and Innovation–Colciencias for the funding of the Ph. D studies of the first author, which include the work presented hereby.

Author details

Leonardo Camargo Forero*, Pablo Royo and Xavier Prats

*Address all correspondence to: leonardo.camargo@upc.edu

Universidad Politécnica de Cataluña, Castelldefels, Spain

References

[1] Message P Forum. Mpi: A Message-Passing Interface Standard. Technical Report. Knoxville, TN, USA: University of Tennessee; 1994

[2] Holland O, Woods J, De Nardi R, Clark A. Beyond swarm intelligence: The ultraswarm. IEEE. In: Proceedings of the IEEE Swarm Intelligence Symposium (SIS '05); 8-10 June 2005. Pasadena, California; 2005, June. pp. 217-224

[3] Marjovi A, Choobdar S, Marques L. Robotic clusters: Multi-robot systems as computer clusters: A topological map merging demonstration. Robotics and Autonomous Systems. 2012;60(9):1191-1204

[4] Dagum L, Menon R. Open MP: An industry standard API for shared-memory programming. IEEE Computational Science and Engineering. 1998;5(1):46-55. DOI: 10.1109/99.660313

[5] Gabriel E, Fagg G, Bosilca G, Angskun T, Dongarra J, Squyres J, Sahay V, Kambadur P, Barrett B, Lumsdaine A, Castain R, Daniel D, Graham R, Woodall T. Open MPI: Goals, concept, and design of a next generation MPI implementation. In: European Parallel Virtual Machine/Message Passing Interface Users' Group Meeting; 19-22 September 2004; Budapest, Hungary: pp. 97-104

[6] Nickolls J, Buck I, Garland M, Skadron K. Scalable parallel programming with CUDA. Queue. 2008;6(2):40-53

[7] Salamí E, Barrado C, Pastor E. UAV flight experiments applied to the remote sensing of vegetated areas. Remote Sensing. 2014;6(11):11051-11081

[8] Intel® Edison Compute Module [Internet]. 2017 Available from: http://www.intel.com/content/www/us/en/do-it-yourself/edison.html [Accessed: 11-01-2017]

[9] NVIDIA Jetson TX1 [Internet]. 2017. Available from: http://www.nvidia.com/object/jetson-tx1-module.html [Accessed: 11-01-2017]

[10] Jeon D, Kim D, Ha Y, Tyan V. Image processing acceleration for intelligent unmanned aerial vehicle on mobile GPU. Soft Computing. 2016:20(5):1713-1720

[11] Cocchioni F, Frontoni E, Ippoliti G, Longhi S, Mancini A, Zingaretti P. Visual based landing for an unmanned quadrotor. Journal of Intelligent & Robotic Systems. 2016: 84(1-4):511-528

[12] Ribeiro C, Dutra M, Rabelo A, Oliveira F, Barbosa A, Frinhani L, Porto D, Milanez R. A robotic flying crane controlled by an embedded computer cluster. In: IEEE 12th Latin American Robotics Symposium (LARS) and 2015 3rd Brazilian Symposium on Robotics (LARS-SBR); 29-31 October 2015; Uberlandia, Brazil: IEEE; 2015. pp. 91-96

[13] ODROID-X2 open development platform [Internet]. 2017. Available from: http://www.hardkernel.com/main/products/prdt_info.php?g_code=G135235611947 [Accessed: 11-01-2017]

[14] Gentzsch W. (Sun Microsystems). Sun grid engine: Towards creating a compute power grid. In: Proceedings of the 1st International Symposium on Cluster Computing and the Grid (CCGRID '01); 16-18 May 2001; Brisnane, Australia: IEEE/ACM; 2001. pp. 35-36

[15] Quigley M, Conley K, Gerkey B, Faust J, Foote T, Leibs J, Wheeler R, Ng A. ROS: An open-source Robot Operating System. In: Proceedings of the ICRA Workshop on Open Source Software; 12-17 May 2009; Kobe, Japan. Vol. 3, No. 3.2. p. 5

[16] Ma'sum M, Jati G, Arrofi M, Wibowo A, Mursanto P, Jatmiko W. Autonomous quadcopter swarm robots for object localization and tracking. In: International Symposium on Micro-NanoMechatronics and Human Science (MHS); 10-13 November 2013; Nagoya, Japan: IEEE; 2013. pp. 1-6

[17] Davis E, Nizette B, Yu C. Development of a low cost quadrotor platform for swarm experiments. In: 32nd Chinese Control Conference (CCC); 26-28 July 2013; Xi'an, China: IEEE; 2013. pp. 7072-7077

[18] Ramaithitima R, Whitzer M, Bhattacharya S, Kumar V. Automated creation of topological maps in unknown environments using a swarm of resource-constrained robots. IEEE Robotics and Automation Letters. 2016;1(2):746-753

[19] Pestana J, Sanchez-Lopez J, de la Puente P, Carrio A, Campoy P. A vision-based quadrotor swarm for the participation in the 2013 international micro air vehicle competition. In: International Conference on Unmanned Aircraft Systems (ICUAS); 27-30 May 2014; Orlando, USA: IEEE; 2014. pp. 617-622

[20] Arumugam R, Enti V, Bingbing L, Xiaojun W, Baskaran K, Kong F, Kumar A, Meng K, Kit G Davinci: A cloud computing framework for service robots. In: IEEE International Conference on Robotics and Automation (ICRA); 3-8 May 2010; Anchorage, USA: IEEE; 2010. pp. 3084-3089

[21] Remmersmann T, Tiderko A, Langerwisch M, Thamke S, Ax M. Commanding multi-robot systems with robot operating system using battle management language. In: IEEE Communications and Information Systems Military Conference (MCC); 8-9 October 2012; Gdansk, Poland: IEEE; 2012. pp. 1-6

[22] Hu C, Hu C, He D, Gu Q. A new ROS-based hybrid architecture for heterogeneous multi-robot systems. In: IEEE 27th Chinese Control and Decision Conference (CCDC); 23-25 May 2015; Qingdao, China: IEEE; 2015. pp. 4721-4726

[23] Raspberry PI 3 model [Internet]. 2017. Available from: https://www.raspberrypi.org/products/raspberry-pi-3-model-b/ [Accessed: 11-01-2017]

[24] ROS documentation concepts [Internet]. 2017. Available from: http://wiki.ros.org/ROS/Concepts [Accessed: 11-01-2017]

Chapter 8

Unmanned Aerial Systems (UASs) for Environmental Monitoring: A Review with Applications in Coastal Habitats

Daniele Ventura, Andrea Bonifazi,
Maria Flavia Gravina and Gian Domenico Ardizzone

Additional information is available at the end of the chapter

http://dx.doi.org/10.5772/intechopen.69598

Abstract

Nowadays the proliferation of small unmanned aerial systems or vehicles (UAS/Vs), formerly known as drones, coupled with an increasing interest in tools for environmental monitoring, have led to an exponential use of these unmanned aerial platforms for many applications in the most diverse fields of science. In particular, ecologists require data collected at appropriate spatial and temporal resolutions to describe ecological processes. For these reasons, we are witnessing the proliferation of UAV-based remote sensing techniques because they provide new perspectives on ecological phenomena that would otherwise be difficult to study. Therefore, we propose a brief review regarding the emerging applications of low-cost aerial platforms in the field of environmental sciences such as assessment of vegetation dynamics and forests biodiversity, wildlife research and management, map changes in freshwater marshes, river habitat mapping, and conservation and monitoring programs. In addition, we describe two applications of habitat mapping from UAS-based imagery, along the Central Mediterranean coasts, as study cases: (1) The upper limit of a *Posidonia oceanica* meadow was mapped to detect impacted areas, (2) high-resolution orthomosaic was used for supporting underwater visual census data in order to visualize juvenile fish densities and microhabitat use in four shallow coastal nurseries.

Keywords: UAVs, UASs, drones, environmental monitoring, coastal habitats, mapping, aerial photography, photogrammetry, image classification, remote sensing, *Posidonia oceanica, Diplodus sargus*, juvenile sparid fish, Mediterranean Sea, Giglio Island

INTECH
open science | open minds

© 2017 The Author(s). Licensee InTech. This chapter is distributed under the terms of the Creative Commons Attribution License (http://creativecommons.org/licenses/by/3.0), which permits unrestricted use, distribution, and reproduction in any medium, provided the original work is properly cited.

1. Introduction

Mankind has always been fascinated by the dream of flight, in fact in many ancient cultures, myths and legends depicted deities with the extraordinary ability to fly like birds. It should be sufficient to recall the Egyptian winged goddess Isis or the Greek myth of Icarus; even Christian iconography preserves and recovers the figures of winged beings as intermediaries between man and God, reinterpreting them as angels. From those ancient times, passing through the Renaissance intuitions of Leonardo Da Vinci, to the first balloon flights of the Montgolfier brothers in 1783, we came to the early twentieth century which witnessed the first sustained and controlled flight of a powered, heavier-than-air machine with a pilot aboard. Just over a hundred of years have passed since that fateful day—December 17, 1903, that is since the Wright Flyer took off near Kill Devil Hills, about four miles south of Kitty Hawk, North Carolina, USA. Nowadays the beauty of flying characterizes our daily lives, becoming an indispensable tool to move people and things in few hours in all parts of the world. We can state that the ability of flight has strongly changed the perspective on our vision of the world.

Long before this first powered flight of Wright brothers, one of the first recorded usages of unmanned aircraft systems (UAS) was by Austrians on August 22, 1849. They launched 200 pilotless balloons, carrying 33 pounds of explosives and armed with half-hour time fuses, against the city of Venice. On May 6, 1896, Samuel P. Langley's Aërodrome No. 5, a steam-powered pilotless model was flown successfully along the Potomac River near Washington. During World War I and World War II, radio-controlled aircrafts were used extensively for aerial surveillance, for training antiaircraft gunners, and they also served as aerial torpedoes (e.g. in 1917, the Hewitt Sperry Automated Airplane, developed by Elmer Sperry, was the early version of today's aerial torpedo). During Cold War, the drone was seen as a viable surveillance platform able to capture intelligence in denied areas. Reconnaissance UASs were first deployed on a large scale in the Vietnam War. By the dawn of the twenty-first century, unmanned aircraft systems were used more and more frequently for a variety of missions especially since the war on terror, becoming a lethal hunter-killer. Due to these historical aspects the public perception of most of the UAV applications is still mainly associated with military use, but nowadays the drone concept is refashioned as a new promise for citizen-led applications having several functions, ranging from monitoring climate change to carrying out search operations after natural disasters, photography, filming, and ecological research.

The interpretation of photos from airborne and satellite-based imagery has become one of the most popular tools for mapping vast surfaces, playing a pivotal role in habitat mapping, measuring, and counting performed in ecological research, as well as to perform environmental monitoring concerning land-use change [1]. However, both satellite and airborne imagery techniques have some disadvantages. For example, the limitations of piloted aircrafts must be considered in regard to their reliance on weather conditions, flight altitude, and speed that can affect the possibility to use such method [1]. In addition, satellite high-resolution data might not be accessible for many developing-country researchers due to financial constraints. Furthermore, some areas such as humid biotopes and tropic coasts are often obscured by a persistent cloud cover, mostly making cloud-free satellite images unavailable for a specific time period and location; moreover the temporal resolution is limited by the availability of aircraft platforms and orbit characteristics of satellites [2]. In addition, the highest spatial resolution data, available from satellites and manned aircraft, is typically in the range of 30–50 cm/pixel. Indeed, for the purpose of monitoring highly dynamic and heterogeneous environments, or for real-time monitoring of land-use change in sensitive habitats, satellite sensors are often limited due to unfavorable revisit times (e.g. 18 days for Landsat) and spatial resolution (e.g. Landsat and Modis ~30 m/pixel) [3]. To address these limitations, new satellite sensors (Quickbird, Pleiades-1A, IKONOS, GeoEye-1, WorldView-3) have become operational over the past decade, offering data at finer than 10-m spatial resolution. Such data can be used for ecological studies, but hurdles such as high cost per scene, temporal coverage, and cloud contamination remain [4].

Emerging from a military background, there is now a proliferation of small civilian unmanned aerial systems or vehicles (UAS/Vs), formerly known as drones. Modern technological advances such as long-range transmitters, increasingly miniaturized components for navigation and positioning, and enhanced imaging sensors have led to an upsurge in the availability of unmanned aerial platforms both for recreational and professional uses. These emerging technologies may provide unprecedented scientific application in the most diverse fields of science. In particular, UAVs offer ecologists a new way to responsive, timely, and cost-effective monitoring of environmental phenomena, allowing the study of individual organisms and their spatiotemporal dynamics at close range [4].

Two main categories of unmanned aerial vehicles (UAVs) exist: rotor-based copter systems and fixed-wing platforms. Rotor-wing units have hovering and VTOL (Vertical Take-Off and Landing) capabilities, while fixed-wing units tend to have longer flight durations and range. However, a more detailed classification can be made according to size, operating range, operational flight altitude, and duration (**Table 1**). For additional information regarding the classification of UAVs, please refer to Refs. [4–6].

Size	Nomenclature	Specifics	Operational requirements	Application areas	Examples
Very large (3–8 tons)	HALE (High Altitude, Long Endurance)	Fly at the highest altitude (> 20 Km) with huge operating range that extend thousands of km, long flight time (over 2 days), very heavy payload capacity (more than 900 kg in under-wing pods)	Prohibitively expensive for most users (high maintenance, sensors, crew training costs), long runway for takeoff and landing, ground-station support, and continuous air-traffic control issues, challenging deployment/recovery and transport	Assessments of climate variable impacts at global scales, remote sensing collection, and earth/ atmospheric science investigations	Global Hawk, Qinetiq Zephyr, NASA PathFinder

Size	Nomenclature	Specifics	Operational requirements	Application areas	Examples
Large (1–3 tons)	MALE (Medium Altitude, Long Endurance)	Medium altitude (3–9 Km), over 12 h flight time with broad operating range (> 500 km), heavy payload capacity (~100 kg internally, external loads of 45 up to 900 kg)	Similar requirements as for HALE but with reduced overall costs	Near-real-time wildfire mapping and surveillance, investigation of storm electrical activity and storm morphology, remote sensing and atmospheric sampling, arctic surveys, atmospheric composition and chemistry	NASA Altus II, NASA Altair, NASA Ikhana, MQ-9 Reaper (Predator B), Heron 2, NASA SIERRA
Medium (25–150 kg)	LALE (Low Altitude, Long Endurance), LASE (Low Altitude, Short Endurance)	Fly at moderate altitude (1–3 Km) with operating ranges that extend from 5 to 150 km), flight time (over 10 hours), moderate payload capacity (10–50 kg)	Reduced costs and requirements for takeoff and landing compared to MALE (hand-launched platforms and catapult-launch platforms), simplified ground-control stations	Remote sensing, mapping, surveillance and security, land cover characterization, agriculture and ecosystem assessment, disaster response and assessment	ScanEagle, Heron 1, RQ-11 Raven, RQ-2 Pionee, RQ-14 Dragon Eye, NASA J-FLiC, Arcturus T-20
Small, mini, and nano (Less than 25 kg for small AUVs, up to 5 Kg for mini and less than 5 Kg for nano)	MAV (Micro) or NAV (Nano) Air Vehicles	Fly at low altitude (< 300 m), with short duration of flight (5–30 min) and range (< 10 Km), small payload capacity (< 5 kg)	Low costs and minimal take off/landing requirements (Hand-launched), often are accompanied by ground-control stations consisting of laptop computers, flown by flight planning software or by direct RC (Visual Line Of Sight or Beyond Visual Line Of Sight when allowed), usually fixed-wing (small AUVs) and copter-type (mini and nano AUVs)	Aerial photography and video, remote sensing, vegetation dynamics, disaster response and assessment, precision agriculture, forestry monitoring, geophysical surveying, photogrammetry, archeological research, environmental monitoring	AR-Parrot, BAT-3, SenseFly eBee, DJI Inspire 3, DJI Phantom 4, Draganflyer X6, Walkera Voyager 4

Table 1. Summary of f UAVs' classes with examples.

In this brief review and in our case studies, we only discuss and illustrate the use of small and mini UAVs because these portable and cost-effective platforms have shown a great potential to deliver high-quality spatial data to a range of science end users.

2. Some recent ecological applications of lightweight UASs

Although lightweight UASs represent only a small fraction of the full list of unmanned systems capable of performing the so-called "three Ds" (i.e. dull, dirty, or dangerous missions), they have been used in a broad range of ecological studies.

2.1. Forest monitoring and vegetation dynamics

Tropical forests play a critical role in the global carbon cycle and harbor around two-thirds of all known species [7]. Tropical deforestation is a major contributor to biodiversity loss, so an urgent challenge for conservationists is to be able to accurately assess and monitor changes in forests, including near real-time mapping of land cover, monitoring of illegal forest activities, and surveying species distributions and population dynamics [2].

Koh and Wich [2] provided with a simple RC fixed-wing UAV (Hobbyking Bixler 2) helpful data for the monitoring of tropical forests of Gunung Leuser National Park in Sumatra, Indonesia. In fact, the acquired images allowed the detection of different land uses, including oil palm plantations, maize fields, logged areas, and forest trails.

UAVs have also been used for the successful monitoring of streams and riparian restoration projects in inaccessible areas on Chalk Creek near Coalville (Utah), as well as to perform non-destructive, nonobtrusive sampling of Dwarf bear claw poppy (*Arctomecon humilis*), a short-lived perennial herb of crust community which is very sensitive to off-road vehicle (ORV) traffic [8]. A fixed-wing (eBee, senseFly) and a quadcopter (Phantom 2 Vision+, DJI) were used [9] to acquire high-spatial resolution photos of an impounded freshwater marsh, demonstrating that UAVs can provide a time-sensitive, flexible, and affordable option to capture dynamic seasonal changes in wetlands, in order to collect effective data for determining percent cover of floating and emergent vegetation.

Dryland ecosystems provide ecosystem services (e.g. food, but also water and biofuel) that directly support 2.4 billion people, covering 40% of the terrestrial area, they characteristically have distinct vegetation structures that are strongly linked to their function [10, 11]. For these reasons, Cunliffe et al. [10] acquired aerial photographs using a 3D Robotics Y6 hexacopter equipped with a global navigation satellite system (GNSS) receiver and consumer-grade digital camera (Canon S100). Later, they processed these images using structure-from-motion (SfM) photogrammetry in order to produce three-dimensional models, describing the vegetation structure of these semi-arid ecosystems. This approach yielded ultrafine (<1 cm^2) spatial resolution canopy height models over landscape-levels (10 ha). This study demonstrated how ecosystem biotic structures can be efficiently characterized at cm scales to process aerial photographs captured from inexpensive lightweight UAS, providing an appreciable advance in the tools available to ecologists. Getzin et al. [12] demonstrated how fine spatial resolution photography (7-cm pixel size) of canopy gaps, acquired with the fixed-wing UAV 'Carolo P200,' can be used to assess floristic biodiversity of the forest understory. Also in riparian contexts, UAS technology provides a useful tool to quantify riparian terrain, to characterize riparian vegetation, and to identify standing dead wood and canopy mortality, as demonstrated by Dunford et al. [13].

2.2. Wildlife research

Often population ecology requires time-series and accurate spatial information regarding habitats and species distribution. UASs can provide an effective means of obtaining such kind of information. Jones et al. [14] used a 1.5-m wingspan UAV equipped with autonomous control system to capture high-quality, progressive-scan video of a number of landscapes and wildlife species (e.g. *Eudocimus albus, Alligator mississippiensis, Trichechus manatus*). Israel [15] dealt with the problem of mortally injured roe deer fawns (*Capreolus capreolus*) by mowing machinery, and demonstrated a technical sophisticated 'detection and carry away' solution to avoid these accidents. In fact, he presented a UAV-based (octocopter Falcon-8 from Ascending Technologies) remote sensing system via thermal imaging for the detection of fawns in the meadows.

Considering that in butterflies, imagoes and their larvae often demand specific and diverging microhabitat structures and resources, Habel et al. [16] took high-resolution aerial images using a DJI Phantom 2 equipped with a H4-3D Zenmuse gimbal and a lightweight digital action camera (GoPro HERO 4). These aerial pictures, coupled with the information on the larvae′s habitat preference from field observations, were used to develop a habitat suitability model to identify preferential microhabitat of two butterfly larvae inhabiting calcareous grassland.

Moreover, UAVs may offer advantages to study marine mammals, in fact Koski et al. [17] used the Insight A-20 equipped with an Alticam 400 (a camera model developed for the ScanEagle UAV) to successfully detect simulated Whale-Like targets, demonstrating the values of such methodology for performing marine ecological surveys. In a similar manner, Hodgson et al. [18] captured 6243 aerial images in Shark Bay (western Australia) with a ScanEagle UVS, equipped with a digital SLR camera, in which 627 containing dugongs, underlying that UAS systems may not be limited by sea state conditions in the same manner as sightings from manned surveys. Whitehead et al. [19] described efforts to map the annual sockeye salmon run along the Adam's River in southern British Columbia, providing an overview of salmon locations through high-resolution images acquired with a lightweight fixed-wing UAV.

3. Case studies along temperate Mediterranean coasts

Although over the past decade there has been an increasing interest in tools for ecological applications such as ultrahigh-resolution imagery acquired by small UAVs, few have been used for environmental monitoring and classification of marine coastal habitats. Indeed, in this section we outline two case studies regarding the application of a small UAV for mapping coastal habitats. These applications represent a cross section of the types of applications for which small UAVs are well-suited, especially when one considers the ecological aspect related to marine species biology and habitat monitoring. Despite the fact that there are a number of advanced sensors that have been developed and many proposed applications for small UASs, here we carried out our studies using a commercially available and low-cost camera.

http://dx.doi.org/10.5772/intechopen.69598

As such, it can be considered a simple, inexpensive, and replicable tool that can be easily implemented in future research which could also be carried out by nonexperts in the field of UASs technologies.

For each survey we used a modified rotary-wing Platform (Quanum Nova CX-20, **Figure 1**), which included an integrated autopilot system (APM v2.5) based on the 'ArduPilot Mega' (APM, http://www.ardupilot.co.uk/), which has been developed by an online community (diydrones.com). The APM includes a computer processor, geographic positioning system module (Ublox Neo-6 Gps), data logger with an inertial measurement unit (IMU), pressure and temperature sensor, airspeed sensor, triple-axis gyro, and accelerometer (**Figure 2**). This quadcopter is relatively inexpensive (<$500) and lightweight (~1.5 Kg). The cameras used to acquire the imagery was a consumer-grade RGB, FULL-HD action camera (Gopro Hero 3 Black Edition, sensor: Complementary Metal-Oxide-Semiconductor; sensor size: 1/2.3″ (6.17 × 4.55 mm), pixel size: 1.55 μm; focal length: 2.77 mm). In addition, a brushless 3-Axis Camera Gimbal (Quanum Q-3D) was installed, to ensure a good stabilization on acquired images, avoiding motion blur. Both drone and gimbal were powered by a ZIPPY 4000 mAh (14.8 V) 4S 25C Lipo battery which allowed a maximum flying time of about 13 min or less, depending on wind. In addition, by combining the APM with the open-source mission planner software (APM Planner), the drone can perform autonomous fly paths and survey grids.

Figure 1. The Quanum Nova CX-20 Quadcopter ready to fly just before a coastal mapping mission.

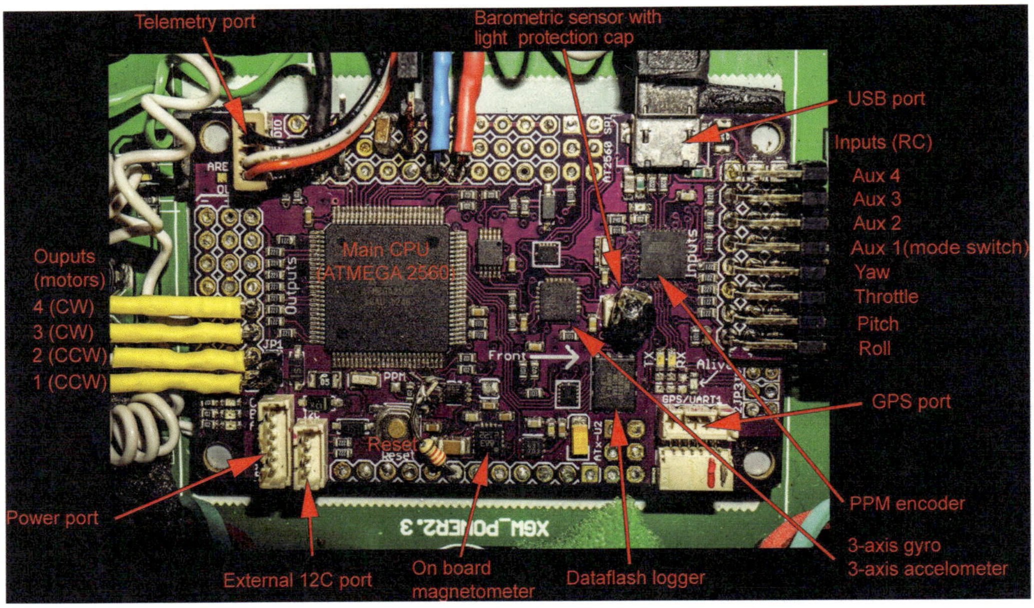

Figure 2. The integrated navigation and autopilot systems (APM 2.5).

3.1. Case study 1: mapping of the upper limit of a *Posidonia oceanica* meadow for the detection of impacted areas

The marine phanerogam *P. oceanica* (L.) Delile is the most widespread seagrass in the Mediterranean Sea [20]. It plays a pivotal role in the ecosystems of shallow coastal waters in several ways by (i) providing habitat for juvenile stages of commercially important species [21]; (ii) significantly reducing coastal erosion, promoting the deposition of particles with dense leaf canopy and thick root-rhizome ('matte') [22]; and (iii) offering a nursery area for many fish and invertebrate species [23, 24]. Although known to be a reef-building organism capable of long-term sediment retention, *P. oceanica* meadows are however experiencing a steep decline throughout the Mediterranean Sea [25, 26]. Along the Mediterranean coasts, the decline of seagrasses on a large spatial scale has been attributed to anthropogenic disturbances such as illegal trawling [27], fish farming [28], construction of marinas [29], and sewage discharge and pollution [30]. On contrast, on a smaller spatial scale, particularly in coastal areas subjected to intense recreational activity, seagrasses are impacted by mechanical damage caused by boat anchoring or moorings [20, 31]. Major damage to seagrasses seems to be caused by dragging anchors and scraping anchor chains along the bottom, as boats swing back and forth, generally resulting in dislodgement of plant rhizomes or leaves [20]. In most published works the mapping of *P. oceanica* meadows has been based on satellite, airborne imagery, multibeam bathymetry, and side-scan sonar mosaics [32]. Remote sensing data from satellites and piloted aircraft can be used to map large areas, but they either do not have adequate spatial resolution or are too expensive to map fine-scale features, otherwise small UAVs are particularly well-suited to mapping the upper limits of meadows at a smaller spatial scale (i.e. 1–5 Km).

The case study for this application was carried out along a sandy cove (Arenella bay) with a well-established *P. oceanica* meadow, approximately 2 km north of Giglio Porto (Giglio Island, Tuscany, IT), in late November 2016. Our goals were to show the high level of detail that can be reached with UAV-based imagery, to respect other free-available remote sensing techniques, and to detect impacted areas of the meadow. In fact, in this study site, there are two principal sources of disturbance: a direct adverse effect on meadow due to boat anchoring during summer seasons, and the presence of a granite quarry that in the past (no longer operational) may have caused an increase in sedimentation rates, resulting in a reduction of cover and shoot density.

We set the GoPro Hero 3 camera to take photos every 2 s (time lapse mode) in Medium Field of View (M FOV: 7 Megapixel format, 3000 × 2250 pixels), and we set the camera pointing 90° downward with auto white balance. Flight speeds were maintained between 5 and 7 m/s to allow for 75% in-track overlap. The drone was programmed to fly at 30 m above mean sea level in order to get a Ground Sampling Distance (GSD) of ~2.5 cm per pixel, according to the formula :

$$GSD_{cm/pix} = \frac{SW_{mm} * FH_{m}}{FL_{mm} * IMW_{pix}} * 100 \quad (1)$$

where GSD is the ground sample distance (i.e. photo resolution on the ground), SW is the sensor width, FH is the flight height, FL is the focal length of the camera, and IMW is the image width. By multiplying the GSD by image size (width and height) the resulting photo footprint was 66 × 50 m.

The bay (1.96 ha) was flown in 16 strips with a total flight duration of 6.34 min. In total, the survey yielded 184 images, which were processed in Adobe Photoshop Lightroom 5.0 (Adobe Systems Incorporated, San Jose, California, USA) using the lens correction algorithm for the GoPro HERO 3 Black Edition camera, in order to remove lens distortion (fish-eye effect). Since for this application, high-spatial accuracy was not required, five ground-control points (GCPs) were placed at accessible locations along the coast (with easily recognizable natural features such as rocks), and they were surveyed with a handheld GPS + GLONASS receiver (Garmin Etrex 30), leading to horizontal errors of ±5 m. Successively, the images where used to produce a high-resolution orthoimage mosaic in Agisoft Photoscan 1.0 (www.Agisoft.com). This structure from motion (SfM) package allows a high degree of automation, and makes it possible for nonspecialists to produce accurate orthophoto mosaics in less time than what it would take using conventional photogrammetric software [19].

Figure 3 shows how high-spatial resolution of RGB imagery acquired from UAV has allowed us to detect the impacted areas of the meadow. In particular, we identified 1.437 m^2 of dead 'matte' by analyzing satellite imagery (Google Earth), 1.686 m^2 with Bing Aerial orthophotos and 1.711 m^2 with UAV-based orthomosaic. In fact, due to the higher spatial resolution of UAS imagery, we were able to detect even the smallest areas where dead 'matte' was exposed, due to meadow degradation (**Figure 3**).

The imagery acquired provides a new perspective on *P. oceanica* mapping and clearly shows how comparative measurements and low-cost monitoring can be made in shallow coastal

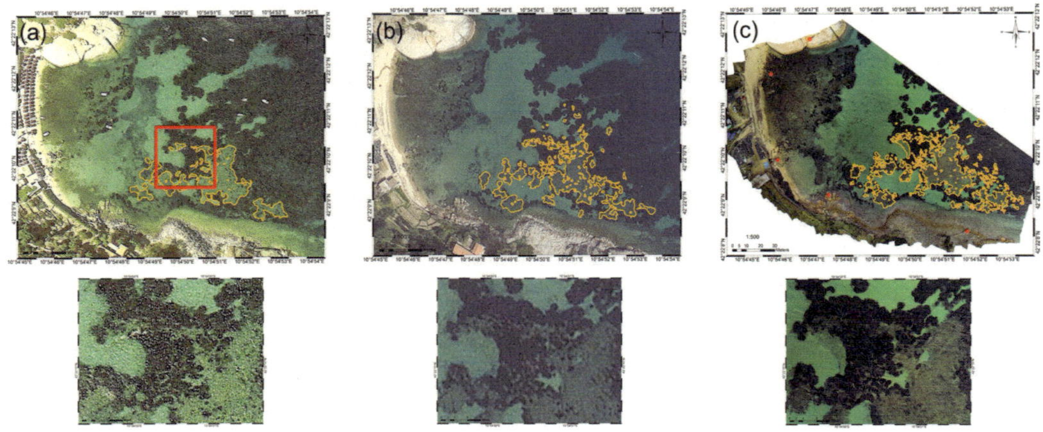

Figure 3. The bay of Arenella (Giglio Island, scale 1:500) with impacted *Posidonia oceanica* meadow (dead *P. oceanica* 'matte' is enclosed by orange polygons) mapped using three different free/low- cost remote sensing techniques: (a) Google Earth Satellite image; (b) Bing aerial orthoimage; and (c) UAV-based orthomosaic. The enclosed area highlighted by the red box is shown at greater scale (1:100) in order to visualize the increasing level of detail. In (c), red dots represent the position of GCPs.

areas. In fact, in this kind of environment, anthropogenic drivers such as boat mooring and creation of coastal dumping areas are significantly affecting ecosystem structure and function. In addition, considering that drone surveying is relatively not expensive, regular time-series monitoring can be adopted to assess the evolution of coastal meadows.

3.2. Case study 2: integration of underwater visual census (UVC) data with UAV-based aerial maps for the characterization of juvenile fish nursery habitats

Most demersal fishes have complex life cycles, in which the adult life-stage takes place in open deeper waters, while juvenile life-stages occur in benthic inshore habitats [33, 34]. The presence of suitable habitats becomes an essential requirement during the settlement of juvenile stages. In fact, these habitats are the key to success for the conclusion of early life phases, providing shelter from predators and abundance of trophic resources. As a result of this site-attachment, juveniles exhibit systematic patterns of distribution, influenced by the availability of microhabitats [33–35]. Habitat identification has been generally achieved by human underwater visual censuses (UVC) techniques [36]. The latter has been considerably improved in recent years with visual underwater video technologies [35]. However, these studies require a deep knowledge of the environment in addition to considerable efforts in terms of time and experienced staff [35]. Small UASs potentially offer a low-cost support to conventional UVC techniques, providing a time-saving tool aimed at improving data from underwater surveys. Indeed, our aim is to couple UVC data (e.g. number of juvenile fish) with remote sensing data (high-resolution UAV-based imagery), to extrapolate habitat features from image analysis, allowing a considerable saving of both time and efforts, especially for underwater operators.

The case study for this application involved the same UAS used in the previous example, an underwater observer, and was focused on a common coastal fish species: the white seabream

(*Diplodus sargus*, L.). *D. sargus* is abundant in the Mediterranean and dominates fish assemblages in shallow rocky infralittoral habitats. It inhabits rocky bottoms and *P. oceanica* beds, from the surface to a depth of 50 m. In common with other sparid fishes, it is an economically important species of interest for fisheries and aquaculture.

Between early May and late June 2016, juvenile white seabream (*D. sargus*, L.) were censused from Cannelle Beach to Cape Marino, along a rocky shoreline (~1.5 km long) south of Giglio Porto (Giglio Island, IT). Counts of fish were obtained from two visual census surveys per month: the diver swam slowly along the shoreline (from 0 to 6 m depth) and recorded the numbers of individuals encountered while snorkeling. When juvenile fish or shoals of settlers (size range 10–55 mm) were observed, the abundance and size of each species were recorded on a plastic slate. In addition, the diver towed a rigid marker buoy with a handheld GPS unit with WAAS correction (GpsMap 62stc) in order to accurately record the position of each shoal of fish.

Two mapping missions were successfully carried out in late July 2016, along the same shoreline, in order to produce a high-resolution aerial map of the coast (**Figure 4**).

The quadcopter flew at 40 m, yielding a ground resolution of ~2.5 cm/pix. The two surveys covered 1446 m of shoreline and took approximately 16 min, resulting in 204 images. Since many stretches of the coast were inaccessible areas, where GCPs cannot be physically measured on

Figure 4. The high-resolution (2.5 cm/pix) mosaic representing the rocky coast (~1.5 Km) south of Cannelle Beach (Giglio Island, Italy), derived from two mapping mission (204 images) of Quanum Nova CX-20.

the ground, we used a direct georeferencing approach. The GPS coordinates of the cameras are determined using the UAV onboard GPS receiver, so that the GPS position at the moment of shot can be written to the EXIF header information for each image, after estimating time offset with Mission Planner (v.1.3.3 or higher) geotagging images tool (for better results preflight synchronization of the camera's internal clock with GPS time is recommended). In addition, these measured values (from onboard GPS) may be useful to estimate the camera's approximate external orientation parameters to speed up photogrammetric workflow (bundle adjustment) in Agisoft Photoscan. However, since they are typically captured at relatively low accuracy in the case of UAVs' consumer-grade GPS, we also registered the final orthomosaics, by importing it as raster image (TIFF format) into Arcmap 10.1 [37]. We aligned the raster with an already existing 1:5000 scale aerial orthophoto by 8 control points in order to perform a 2nd order polynomial transformation. Afterward, the control points were used to check the reliability of image transformations. The total error was computed by taking the root mean square sum of all the residuals to compute the RMS error (RMSE). This value described how consistent the transformation was between the different control points. The RMSE achieved was 0.15 pixels which was well under the conventional requirements of less than 1 pixel [38]. The successful geo-registration allowed a direct visualization on the map of UVC data (i.e. lat/long coordinates of fish shoals) after downloading GPS eXchange (.gpx) information from GPS unit. These GPS data were imported as point shapefile in ArcMap using DNRGPS 6.1.0.6 application [39].

As all juvenile fish positions, with their relative abundances (number of fish per shoal), are now available in a GIS environment, it is a straightforward process to model them with interpolation methods, which is, for example, available in ArcMap. The point data, measured from irregularly spaced locations, were converted into continuous surfaces using an inverse distance weighting (IDW) method and then rasterized into a grid format. We used local interpolators of inverse distance weighting because the concept of computation (i.e. it assumes that each point has a local influence that diminishes with distance [40]) is relevant for juvenile fish, where closer points are thought to be similar as a result of the habitat characteristics. **Figures 5** and **6** show the spatial distribution of *D. sargus* juvenile density collected through underwater visual census after IDW interpolation. GIS data integration allowed us to identify two important aspects: (1) four areas (a–d) with high densities of juveniles were clearly visible, suggesting that such zones serve as nursery grounds for juvenile white seabreams (**Figure 5**) and (2) as the juveniles grew larger in size (> 40 mm) a dispersal out of the nursery areas was evident and the preference for a given habitat type decreased leading to an increase in the number of shoals but with lower densities within shoals (**Figure 6**).

These four nurseries were investigated through image analysis (**Figure 7a–d**): we performed a Maximum Likelihood Classification algorithm followed by both postprocessing workflow [41] and manual polygons editing for edge refinement in order to highlight the most important habitat feature such as substrata type and extent (**Table 2**). In fact, due to high site-attachment of juvenile fish, the presence of specific habitats play a key role in the development of early life-stages, hence the fine characterization of these environments becomes an important aspect regarding ecological studies focused on juvenile fish. However, underwater data collection by SCUBA operators require a large effort to acquire such detailed information, therefore UASs-based remote sensing techniques become useful, reliable, and feasible tools for mapping coastal fish habitats and for supporting ecological investigations.

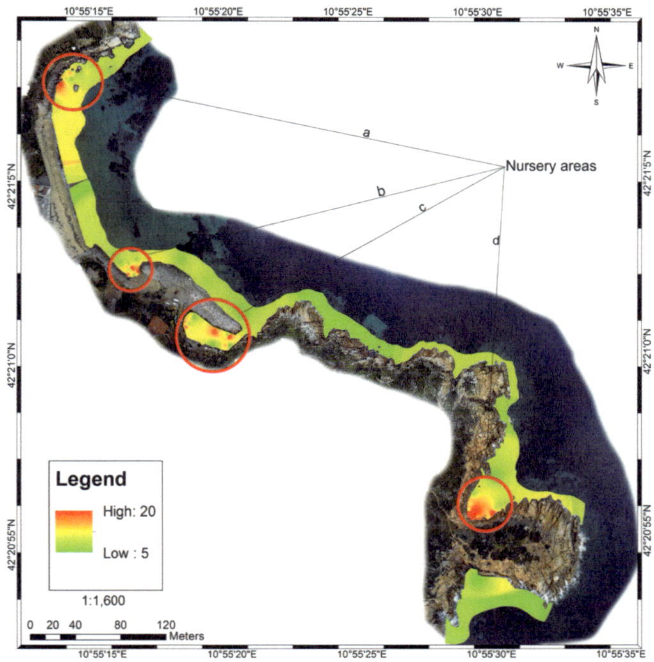

Figure 5. Spatial distribution of small-sized (10–40 mm) juvenile *D. sargus*. IDW-interpolated fish density after UVC data collection in May 2016. The four areas (a–d) with the highest densities of juvenile are highlighted by red circles.

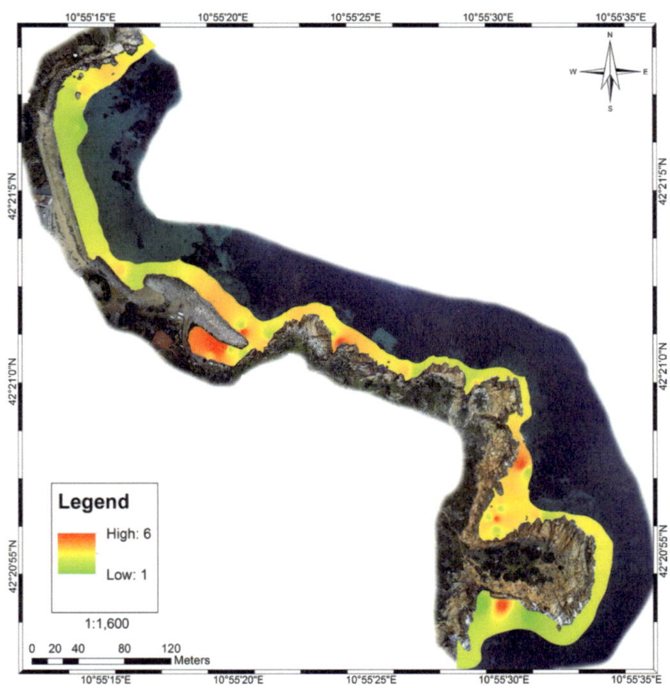

Figure 6. Spatial distribution of large-sized (41–55 mm) juvenile *D. sargus*. IDW-interpolated density after UVC data collection in late June 2016.

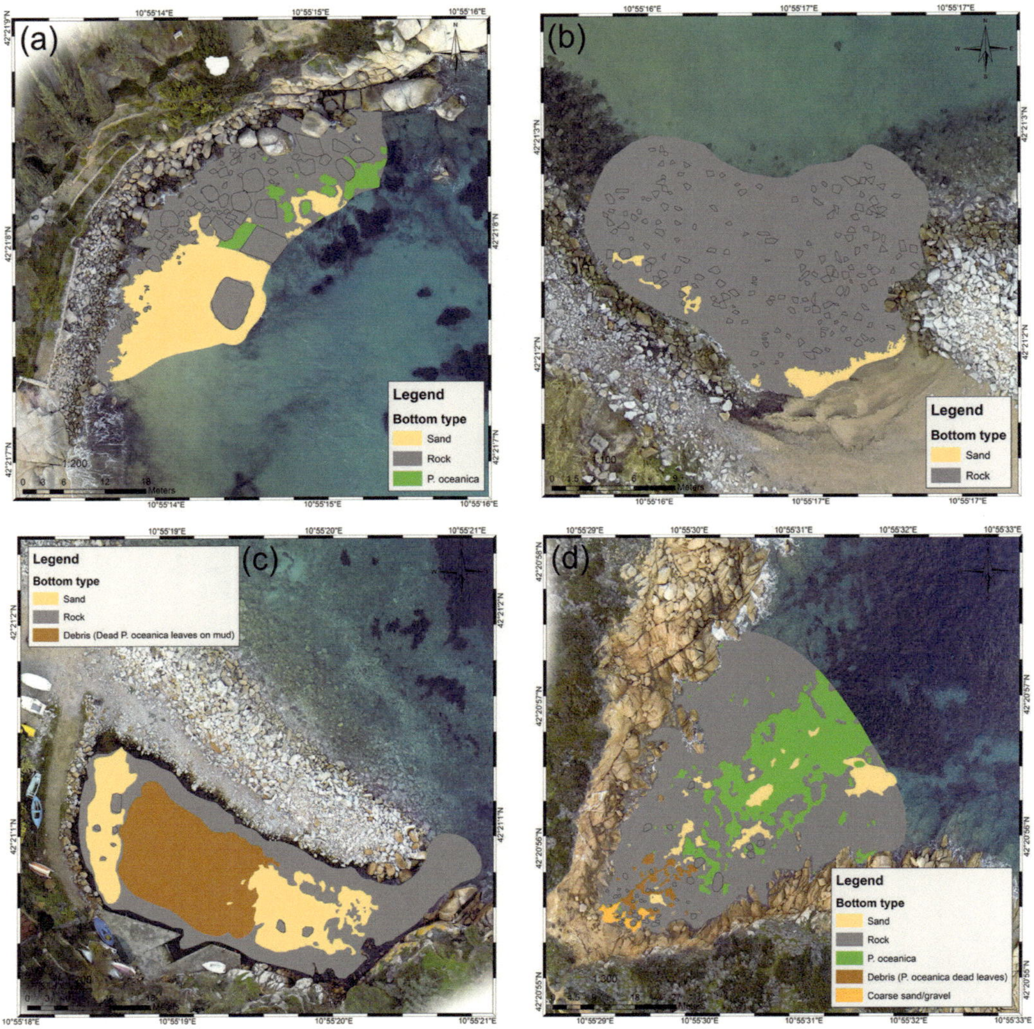

Figure 7. Thematic maps of the four nursery areas (a–d) derived from Maximum likelihood classification and manual editing. Different colors represent main habitat types.

Nursery	Substrata type	Habitat description	Habitat cover (m²)	Percent cover (%)	Depth (m)	Total extent (m²)
Sandy cove (a)	Sand	Granite coarse sand	243.9	36.2	0–3	674.4
	Rock	Large- (mean ± SD diameter: 3.4 ± 16) and medium-sized (mean ± SD diameter: 0.9 ± 0.3) boulders with photophilic algae biocenosis	382.5	56.7		
	Posidonia oceanica	Small patches on sand	48	7.1		

Nursery	Substrata type	Habitat description	Habitat cover (m^2)	Percent cover (%)	Depth (m)	Total extent (m^2)
Rocky cove (b)	Sand	Granite coarse sand	11.6	3.7	0–3.5	317.8
	Rock	Small-sized (mean ± SD diameter: 0.6 ± 0.2) blocks and pebbles with photophilic algae biocenosis	306.2	96.3		
Small port (c)	Sand	Fine sand and mud	218.7	23.8	0–2.8	918.3
	Rock	Cranny rock semisciaphilic algae and isolated boulders on soft sediment	402.2	43.8		
	Debris	Dead *P. oceanica* leaves on mud	297.4	32.4		
Rocky/sandy cove (d)	Sand	Sandy patches	129.1	5.1	0–5.5	2521.2
	Gravel and pebbles	Small- and medium-sized pebbles on sand	25.2	1		
	Rock	Cranny rock with photophilic algae biocenosis and isolated boulders	1698.9	67.4		
	Posidonia oceanica	*Posidonia* meadow and 'matte'	621	24.6		
	Debris	Dead *P. oceanica* leaves on sand	47	1.9		

Measures are derived from high-resolution mosaics image analysis in Arcmap 10.1.

Table 2. Main habitat features characterizing the four nursery areas (a–d).

4. Conclusions

In this brief review we have provided an overview of ecological studies carried out with small drones. Through study cases we demonstrated how UAV-acquired imagery has a substantial potential to revolutionize the study of coastal ecosystem dynamics. The future of UASs applications looks very promising due to the relative low cost with respect to the benefits obtained [42]. In fact, the field of ecology is severely hindered by the difficulties of acquiring appropriate data, and particularly data at fine spatial and temporal resolutions, at reasonable costs [4]. As demonstrated in this study, unmanned aerial vehicles offer ecologists new opportunities for scale-appropriate measurements of ecological phenomena providing land cover information with a very high, user-specified resolution, allowing for fine mapping and characterization of coastal habitats. Although the camera equipment used herein only captures three color (RGB) channels with relatively low resolution (max 16 megapixel), it was possible to distinguish impacted areas in sensitive habitat types, as well as preferred sites for juvenile fish species. Moreover, high-spatial resolution data derived from UAVs combined with traditional underwater visual census techniques enable the direct visualization of field data into geographic space bringing spatial ecology toward new perspectives. High-resolution aerial mosaics allow rapid detection of key habitats, and thus can be used to identify areas of high relevance for species protection and areas where management action should be implemented to improve or maintain habitat quality [16]. UASs are potentially useful to investigate population trends and habitat use patterns, and to assess the effect of human activities (e.g. tourism, pollution) on abundance, particularly in coastal and shallow habitats, where visibility enables animal detection from the surface, as demonstrated for elasmobranch species in coral reef habitats [44]. Finally, although the flexibility of UASs will be able to revolutionize the way we address and solve ecological problems [9], we must consider government approval navigational stipulations and social implications that impose restrictions on the use of UASs before undertaking research projects involving the use of UASs [43–45].

Acknowledgements

The authors thank Dr M. Bruno and Dr A. Belluscio for their assistance during drone's configuration and field sampling, respectively. Acknowledgments are also given to V. Lopetuso and Editors for their valuable comments regarding this manuscript.

Author details

Daniele Ventura[1]*, Andrea Bonifazi[2], Maria Flavia Gravina[2] and Gian Domenico Ardizzone[1]

*Address all correspondence to: daniele.ventura@uniroma1.it

1 Department of Environmental Biology, University of Rome "La Sapienza", Rome, Italy

2 Laboratory of Experimental Ecology and Aquaculture, University of Rome "Tor Vergata", Rome, Italy

References

[1] Kalmár S. Different applications of aerial photography using radio controlled model airplanes. In: International Scientific Conference on Sustainable Development & Ecological Footprint; 26-27 May 2012; Sopron, Hungary. 2012. pp. 1-4

[2] Koh LP, Wich SA. Dawn of drone ecology: Low-cost autonomous aerial vehicles for conservation. Tropical Conservation Science. 2012;5(2):121-132

[3] Turner D, Lucieer A, Watson C. An automated technique for generating georectified mosaics from ultra-high resolution unmanned aerial vehicle (UAV) imagery, based on structure from motion (SfM) point clouds. Remote Sensing. 2012;4(5):1392-1410. DOI: 10.3390/rs4051392

[4] Anderson K, Gaston KJ. Lightweight unmanned aerial vehicles will revolutionize spatial ecology. Frontiers in Ecology and the Environment. 2013;11(3):138-146. DOI: 10.1890/120150

[5] Watts AC, Ambrosia VG, Hinkley EA. Unmanned aircraft systems in remote sensing and scientific research: Classification and considerations of use. Remote Sensing. 2012;4(6):1671-1692. DOI: 10.3390/rs4061671

[6] Baiocchi V, Dominici D, Mormile M. Unmanned aerial vehicle for post seismic and other hazard scenarios. Wit Transactions on the Built Environment. In: Garcia F, Brebbia CA, Guarascio M, editors. Safety and Security Engineering V. Vol. 134. Southampton, Boston: WIT Press; 2013. pp. 113-122. DOI: 10.2495/SAFE130111

[7] Paneque-Gálvez J, McCall MK, Napoletano BM, Wich SA, Koh LP. Small drones for community-based forest monitoring: An assessment of their feasibility and potential in tropical areas. Forests. 2014;5(6):1481-1507. DOI: 10.3390/f5061481

[8] Quilter MC, Anderson VJ. Low altitude/large scale aerial photographs: A tool for range and resource managers. Rangelands Archives. 2000;22(2):13-17. DOI: 10.2458/azu_rangelands_v22i2_Quilter

[9] Marcaccio JV, Markle CE, Chow-Fraser P. Use of fixed-wing and multi-rotor unmanned aerial vehicles to map dynamic changes in a freshwater marsh. Journal of Unmanned Vehicle Systems. 2016;4(3):193-202. DOI: 10.1139/juvs-2015-0016

[10] Cunliffe AM, Brazier RE, Anderson K. Ultra-fine grain landscape-scale quantification of dryland vegetation structure with drone-acquired structure-from-motion photogrammetry. Remote Sensing of Environment. 2016;183:129-143. DOI: 10.1016/j.rse.2016.05.019

[11] Scott RL, Biederman JA, Hamerlynck EP, Barron-Gafford GA. The carbon balance pivot point of Southwestern U.S. semiarid ecosystems: Insights from the 21st century drought. Journal of Geophysical Research: Biogeosciences. 2015;120(12):2612-2624. DOI: 10.1002/2015JG003181

[12] Getzin S, Wiegand K, Schöning I. Assessing biodiversity in forests using very high-resolution images and unmanned aerial vehicles. Methods in Ecology and Evolution. 2012;3(2):397-404. DOI: 10.1111/j.2041-210X.2011.00158.x

[13] Dunford R, Michel K, Gagnage M, Piégay H, Trémelo ML. Potential and constraints of unmanned aerial vehicle technology for the characterization of Mediterranean riparian forest. International Journal of Remote Sensing. 2009;30(19):4915-4935. DOI: 10.1080/01431160903023025

[14] Jones GP IV, Pearlstine LG, Percival HF. An assessment of small unmanned aerial vehicles for wildlife research. Wildlife Society Bulletin. 2006;34(3):750-758. DOI: 10.2193/0091-7648(2006)34[750:AAOSUA]2.0.CO;2

[15] Israel M. A UAV-based roe deer fawn detection system. International Archives of Photogrammetry and Remote Sensing. 2011;38:1-5

[16] Habel JC, Teucher M, Ulrich W, Bauer M, Rödder D. Drones for butterfly conservation: Larval habitat assessment with an unmanned aerial vehicle. Landscape Ecology. 2016;31(10):2385-2395. DOI: 10.1007/s10980-016-0409-3

[17] Koski WR, Allen T, Ireland D, et al. Evaluation of an unmanned airborne system for monitoring marine mammals. Aquatic Mammals. 2009;35(3):347-357. DOI: 10.1578/AM.35.3.2009.347

[18] Hodgson A, Kelly N, Peel D. Unmanned aerial vehicles (UAVs) for surveying marine fauna: A dugong case study. PloS One. 2013;8(11):e79556. DOI: 10.1371/journal.pone.0079556

[19] Whitehead K, Hugenholtz CH, Myshak S, Brown O, LeClair A, Tamminga A, Barchyn TE, Moorman B, Eaton B. Remote sensing of the environment with small unmanned aircraft systems (UASs), part 2: Scientific and commercial applications. Journal of Unmanned Vehicle Systems. 2014;2(3):86-102. DOI: 10.1139/juvs-2014-0006

[20] Milazzo M, Badalamenti F, Ceccherelli G, Chemello R. Boat anchoring on Posidonia oceanica beds in a marine protected area (Italy, Western Mediterranean): Effect of anchor types in different anchoring stages. Journal of Experimental Marine Biology and Ecology. 2004;299(1):51-62. DOI: 10.1016/j.jembe.2003.09.003

[21] Francour P. Fish assemblages of Posidonia oceanica beds at Port-Cros (France, NW Mediterranean): Assessment of composition and long-term fluctuations by visual census. Marine Ecology. 1997;18(2):157-173. DOI: 10.1111/j.1439-0485.1997.tb00434.x

[22] Cavazza W, Immordino F, Moretti L, Peirano A, Pironi A, Ruggiero F. Sedimentological parameters and seagrasses distributions as indicators of anthropogenic coastal degradation at Monterosso Bay (Ligurian Sea, NW Italy). Journal of Coastal Research. 2000;16(2):295-305

[23] Guidetti P. Differences among fish assemblages associated with nearshore Posidonia oceanica seagrass beds, rocky–algal reefs and unvegetated sand habitats in the Adriatic Sea. Estuarine, Coastal and Shelf Science. 2000;50(4):515-529. DOI: 10.1006/ecss.1999.0584

[24] Francour P, Le Direac'h L. Recrutement de l'ichtyofaune dans l'herbier superficiel à Posidonia oceanica dans la réserve marine de Scandola (Corse, Méditerranée nord-occidentale): données préliminaires. Travaux scientifiques du Parc naturel régional et des réserves naturelles de Corse. 1994;46:71-91

[25] Peres JM, Picard J. Causes de la raréfaction et de la disparition des herbiers de Posidonia oceanica sur les côtes françaises de la Méditerranée. Aquatic Botany. 1975;1:133-139. DOI: 10.1016/0304-3770(75)90018-2

[26] Boudouresque CF, Bernard G, Pergent G, Shili A, Verlaque M. Regression of Mediterranean seagrasses caused by natural processes and anthropogenic disturbances and stress: A critical review. Botanica Marina. 2009;52(5):395-418. DOI: 10.1515/BOT.2009.057

[27] Kiparissis S, Fakiris E, Papatheodorou G, Geraga M, Kornaros M, Kapareliotis A, Ferentinos G. Illegal trawling and induced invasive algal spread as collaborative factors in a Posidonia oceanica meadow degradation. Biological Invasions. 2011;13(3):669-678. DOI: 10.1007/s10530-010-9858-9

[28] Ruiz JM, Pérez M, Romero J. Effects of fish farm loadings on seagrass (Posidonia oceanica) distribution, growth and photosynthesis. Marine Pollution Bulletin. 2001;42(9):749-760. DOI: 10.1016/S0025-326X(00)00215-0

[29] Tuya F, Martín, JA, Luque A. Impact of a marina construction on a seagrass bed at Lanzarote (Canary Islands). Journal of Coastal Conservation. 2002;8(2):157-162

[30] Ralph PJ, Tomasko D, Moore K, Seddon S, Macinnis-Ng CM. Human impacts on seagrasses: Eutrophication, sedimentation, and contamination. In: Larkum AWD, Orth RJ, Duarte CM, editors. Seagrasses: Biology, Ecology and Conservation. Dordrecht, The Netherlands: Springer; 2007. pp. 567-593

[31] Walker DI, Lukatelich RJ, Bastyan G, McComb AJ. Effect of boat moorings on seagrass beds near Perth, Western Australia. Aquatic Botany. 1989;36(1):69-77. DOI: 10.1016/0304-3770(89)90092-2

[32] Telesca L, Belluscio A, Criscoli A, Ardizzone GD, Apostolaki ET, Fraschetti S, Gristina M, Knittweis L, Martin CS, Pergent G, Alagna A, Badalamenti F, Garofalo G, Gerakaris V, Pace ML, Pergent-Martin C, Salomidi M. Seagrass meadows (Posidonia oceanica) distribution and trajectories of change. Scientific Reports. 2015;5:1-14. DOI: 10.1038/srep12505

[33] Vigliola L, Harmelin-Vivien ML, Biagi F, Galzin R, Garcia-Rubies A, Harmelin JG, Jouvenel JV, Le Direach L, Macpherson E, Tunesi L. Spatial and temporal patterns of settlement among sparid fishes of the genus Diplodus in the Northwestern Mediterranean. Marine Ecology Progress Series. 1998 ;168:45-56. DOI: 10.3354/meps168045

[34] Macpherson E. Ontogenetic shifts in habitat use and aggregation in juvenile sparid fishes. Journal of Experimental Marine Biology and Ecology. 1998;220(1):127-150. DOI: 10.1016/S0022-0981(97)00086-5

[35] Ventura D, Bruno M, Lasinio GJ, Belluscio A, Ardizzone G. A low-cost drone based application for identifying and mapping of coastal fish nursery grounds. Coastal and Shelf Science. 2016;171:85-98. DOI: 10.1016/j.ecss.2016.01.030

[36] Harmelin-Vivien ML, Harmelin JG, Chauvet C, Duval C, Galzin R, Lejeune P, Barnabe G, Blanc F, Ghevalier R, Duclerc J, Lasserre G. Evaluation visuelle des peuplements et populations de poissons méthodes et problèmes. Revue d'Écologie (La Terre et la Vie). 1985;40(4):468-539

[37] Esri. About ArcGIS [Internet]. Available from: http://www.esri.com/arcgis/about-arcgis [Accessed: January 2017]

[38] Rozenstein O, Karnieli A. A comparison of methods for land-use classification incorporating remote sensing and GIS inputs. Applied Geography. 2011;31(2):533-544. DOI: 10.1016/j.apgeog.2010.11.006

[39] Minnesota Department of Natural Resources. DNRGPS Application [Internet]. 2012 [Updated: 2014]. Available from: http://www.dnr.state.mn.us/mis/gis/DNRGPS/DNRGPS.html [Accessed: January 2, 2017]

[40] Gonzalez A, Marin VH. Distribution and life cycle of Calanus chilensis and Centropages brachiatus (Copepoda) in Chilean coastal waters: A GIS approach. Marine Ecology Progress Series. 1998;165:109-117. DOI: 10.3354/meps165109

[41] Esri ArcGIS. Processing Classified Output [Internet]. [Updated: 2016]. Available from: http://desktop.arcgis.com/en/arcmap/latest/extensions/spatial-analyst/image-classification/processing-classified-output.htm [Accessed: January 17, 2017]

[42] Pajares G. Overview and current status of remote sensing applications based on unmanned aerial vehicles (UAVs). Photogrammetric Engineering & Remote Sensing. 2015;81(4):281-329. DOI: 10.14358/PERS.81.4.281

[43] Kiszka JJ, Mourier J, Gastrich K, Heithaus MR. Using unmanned aerial vehicles (UAVs) to investigate shark and ray densities in a shallow coral lagoon. Marine Ecology Progress Series. 2016;560:237-242. DOI: 10.3354/meps11945

[44] Vincent JB, Werden LK, Ditmer MA. Barriers to adding UAVs to the ecologist's toolbox. Frontiers in Ecology and the Environment. 2015;13(2):74-75. DOI: 10.1890/15.WB.002

[45] Sandbrook C. The social implications of using drones for biodiversity conservation. Ambio. 2015;44(4):636-647. DOI: 10.1007/s13280-015-0714-0